Lecture Notes in Computer Science **14311**

The series Lecture Notes in Computer Science (LNCS), including its subseries Lecture Notes in Artificial Intelligence (LNAI) and Lecture Notes in Bioinformatics (LNBI), has established itself as a medium for the publication of new developments in computer science and information technology research, teaching, and education.

LNCS enjoys close cooperation with the computer science R & D community, the series counts many renowned academics among its volume editors and paper authors, and collaborates with prestigious societies. Its mission is to serve this international community by providing an invaluable service, mainly focused on the publication of conference and workshop proceedings and postproceedings. LNCS commenced publication in 1973.

Andre Esser · Paolo Santini
Editors

Code-Based Cryptography

11th International Workshop, CBCrypto 2023
Lyon, France, April 22–23, 2023
Revised Selected Papers

Springer

Editors
Andre Esser
Technology Innovation Institute
Abu Dhabi, United Arab Emirates

Paolo Santini
Marche Polytechnic University
Ancona, Italy

ISSN 0302-9743 ISSN 1611-3349 (electronic)
Lecture Notes in Computer Science
ISBN 978-3-031-46494-2 ISBN 978-3-031-46495-9 (eBook)
https://doi.org/10.1007/978-3-031-46495-9

This Springer imprint is published by the registered company Springer Nature Switzerland AG
The registered company address is: Gewerbestrasse 11, 6330 Cham, Switzerland

Preface

Post-quantum cryptography has seen tremendous interest in recent years, especially after the call for standardization of quantum-safe primitives by the US National Institute of Standards and Technology (NIST) back in 2017. The first draft PQC standards have just been published, while the competition is still ongoing with a fourth (and probably last) round. NIST has also issued a call for additional post-quantum digital signatures; the first round started earlier this summer.

Code-based cryptography, i.e., the study of schemes based on coding theory assumptions, is arguably among the most relevant and active areas in post-quantum cryptography. As such, all remaining candidates of the NIST PQC fourth round, as well as several proposals submitted to the call for additional signatures, follow code-based assumptions.

Originally named the "Code-Based Cryptography (CBC) Workshop", the series was initiated in 2009 as an informal forum with the goal of bringing together researchers active in the analysis and development of code-based encryption and authentication schemes. Over the years, the workshop has grown from a Europe-based, regional event to become a worldwide venue for the code-based cryptography community. The workshop was renamed "CBCrypto" in 2020, and its organization was co-located with the flagship conference Eurocrypt, and extended to include the publication of revised selected manuscripts in the form of a post-conference proceedings volume. Quickly, CBCrypto has become a popular event with high participation. The 2023 edition of CBCrypto, held in Lyon, France with more than 120 registrations for physical attendance and 80 for online participation, confirmed this trend.

Featuring 8 sessions and 2 invited talks, there were 22 contributed talks over 2 days, presenting recent research and works in progress. This book collects the 8 contributions that were selected for publication by the Program Committee through a careful peer review process. These contributions span many important aspects of code-based cryptography such as cryptanalysis of existing schemes, the proposal of new cryptographic systems and protocols, as well as improved decoding algorithms. As such, the works presented in this book provide a synthesized yet significant overview of the state of the art of code-based cryptography, laying out the groundwork for future developments. We wish to thank the Program Committee members and the external reviewers for their hard and timely work.

September 2023

Andre Esser
Paolo Santini

Organization

General Chairs

Andre Esser	Technology Innovation Institute, UAE
Paolo Santini	Marche Polytechnic University, Italy

Program Committee Chairs

Andre Esser	Technology Innovation Institute, UAE
Paolo Santini	Marche Polytechnic University, Italy

Steering Committee

Marco Baldi	Marche Polytechnic University, Italy
Hannes Bartz	German Aerospace Center, Germany
Gretchen Matthews	Virginia Tech, USA
Edoardo Persichetti	Florida Atlantic University, USA
Joachim Rosenthal	University of Zurich, Switzerland
Antonia Wachter-Zeh	Technical University of Munich, Germany

Program Committee

Marco Baldi	Marche Polytechnic University, Italy
Hannes Bartz	German Aerospace Center, Germany
Emanuele Bellini	Technology Innovation Institute, UAE
Sergey Bezzateev	Saint Petersburg State University of Aerospace Instrumentation, Russia
Loïc Bidoux	Technology Innovation Institute, UAE
Pierre-Louis Cayrel	Laboratoire Hubert Curien, France
Franco Chiaraluce	Marche Polytechnic University, Italy
Jean-Christophe Deneuville	École Nationale de l'Aviation Civile, France
Taraneh Eghlidos	Sharif University of Technology, Iran
Andre Esser	Technology Innovation Institute, UAE
Shay Gueron	University of Haifa and Intel Corporation, Israel
Felice Manganiello	Clemson University, USA

Chiara Marcolla	Technology Innovation Institute, UAE
Giacomo Micheli	University of South Florida, USA
Kirill Morozov	University of North Texas, USA
Gerardo Pelosi	Politecnico di Milano, Italy
Edoardo Persichetti	Florida Atlantic University, USA
Joachim Rosenthal	University of Zurich, Switzerland
Simona Samardjiska	Radbound University, The Netherlands
Paolo Santini	Marche Polytechnic University, Italy
Antonia Wachter-Zeh	Technical University of Munich, Germany
Violetta Weger	Technical University of Munich, Germany
Øyvind Ytrehus	University of Bergen, Norway

Additional Reviewer

Marco Timpanella

Contents

An Analysis of the RankSign Signature Scheme with Rank Multipliers

Anna Baumeister[1,2](✉), Hannes Bartz[2], and Antonia Wachter-Zeh[1]

[1] Technical University of Munich (TUM), Munich, Germany
{anna.baumeister,antonia.wachter-zeh}@tum.de

[2] German Aerospace Center (DLR), Oberpfaffenhofen-Wessling, Germany
hannes.bartz@dlr.de

Abstract. We investigate the application of rank multipliers as introduced by Loidreau to repair the 2017 NIST proposal RankSign. In RankSign, a signature is generated by interpreting the message to be signed as the syndrome of a low-rank parity-check (LRPC) code. Through knowledge of $\mathcal{F}$, the low-dimensional subspace of $\mathbb{F}_{q^m}$ from which the elements of the parity-check matrix are drawn, this syndrome can be decoded to a low-rank signature $\mathbf{e}$. Thus, the security of RankSign crucially relies on the obfuscation of $\mathcal{F}$, ideally making the public code indistinguishable from a random code. Unfortunately, RankSign was broken shortly after its submission by an attack exploiting low-rank codewords in the public code due to the right scrambler being chosen over the base field $\mathbb{F}_q$. We propose to adapt RankSign by using the inverse of a rank multiplier as a right scrambler, increasing the minimum distance of both the public code $\mathcal{C}_{pub}$ and its dual $\mathcal{C}_{pub}^{\perp}$. With this change, the public code contains significantly fewer codewords of low-rank weight, thus preventing the attack that broke the initial RankSign proposal.

Keywords: Code-Based Cryptography · Post-Quantum Cryptography · Rank Metric · Digital Signatures

1 Introduction

This contribution explores the use of rank multipliers introduced by Loidreau [7] to repair the digital signature scheme RankSign [5], a round-1 candidate in the NIST Post-Quantum Cryptography standardization project. RankSign is a hash-and-sign signature scheme whose security relies on the hardness of the rank syndrome decoding (RSD) problem: A signature is generated by interpreting the message to be signed (or rather a fixed-length hash thereof) as the syndrome of a public augmented low-rank parity-check (LRPC) code. This syndrome is then decoded to a low rank-weight signature using knowledge of $\mathcal{F}$, the low dimensional subspace of $\mathbb{F}_{q^m}$ from which the elements of the parity-check matrix $\mathbf{H}$ are drawn.

Since knowledge of $\mathcal{F}$ alone allows to sign, $\mathbf{H}$ must be obfuscated in order to serve as the public key. The most general form of the public key is $\mathbf{H}_{pub} =$

A. Esser and P. Santini (Eds.): CBCrypto 2023, LNCS 14311, pp. 1–13, 2023.
https://doi.org/10.1007/978-3-031-46495-9_1

$\mathbf{Q}[\mathbf{R} \mid \mathbf{H}]\mathbf{P}$, where $\mathbf{H}$ is the parity-check matrix of an LRPC code whose elements are drawn from a d-dimensional subspace $\mathcal{F}$, $\mathbf{Q}$ and $\mathbf{P}$ are invertible matrices and $\mathbf{R}$ is a random matrix.
In the original RankSign proposal [5], the right scrambler $\mathbf{P}$ was chosen as an invertible matrix over the base field $\mathbb{F}_q$, since the transformation by $\mathbf{P}$ is then an isometry for the rank metric and does not change the rank weight of the error. Together with an unfortunate choice of parameters, this resulted in the scheme being broken shortly after its submission by an attack exploiting low-weight codewords in the public code $\mathcal{C}_{pub}$ that were not affected by the matrix $\mathbf{P}$ [2]. In this proposal, we investigate the use of an inverted rank multiplier over the extension field $\mathbb{F}_{q^m}$ as a right scrambler $\mathbf{P}$ to increase the minimum distance of the public code $\mathcal{C}_{pub}$ and its dual $\mathcal{C}_{pub}^{\perp}$. With this adaptation, the public code contains significantly fewer codewords of low-rank weight d, thus preventing the attack by Debris-Alazard and Tillich [2].

2 Rank Metric Properties and Bounds

In this section, we will review the definition of rank-metric codes along with some useful constructions and bounds. Finally, the rank syndrome decoding problem is introduced upon whose security RankSign relies.

2.1 The Rank Metric

Let $\mathbb{F}_q$ be the finite field of order q and $\mathbb{F}_{q^m}$ its extension field of extension order m. Given a vector $\mathbf{v} \in \mathbb{F}_{q^m}^n$ and a basis $\beta = (\beta_1, \ldots, \beta_m)$ of $\mathbb{F}_{q^m}$ over $\mathbb{F}_q$, we can associate to $\mathbf{v}$ a matrix $\mathbf{M_v} \in \mathbb{F}_q^{m \times n}$ where the ith column of $\mathbf{M_v}$ is the representation of v_i in the basis β

$$(v_1, \ldots, v_n) \leftrightarrow \begin{pmatrix} v_{1,1} & \cdots & v_{1,n} \\ \vdots & \ddots & \vdots \\ v_{m,1} & \cdots & v_{m,n} \end{pmatrix}$$

The rank weight of a vector $\mathbf{v}$ is then the rank of its associated matrix $\mathbf{M_v}$:

$$|\mathbf{v}|_r := rank(\mathbf{M_v})$$

and the distance between two vectors $\mathbf{a}, \mathbf{b}$ in the rank metric is:

$$d_r(\mathbf{a}, \mathbf{b}) = rank(\mathbf{M_a} - \mathbf{M_b}).$$

Even though the associated matrix $\mathbf{M_v}$ depends on the choice of basis β for $\mathbb{F}_{q^m}$, the rank of $\mathbf{M_v}$ and thus the rank weight of the associated vector does not.

In the context of the rank metric, we often need to find the product space $\langle \mathcal{AB} \rangle$ of two $\mathbb{F}_q$-subspaces $\mathcal{A}$ and $\mathcal{B}$ of $\mathbb{F}_{q^m}$. It is the space generated by all $\mathbb{F}_{q^m}$-linear combinations of elements in $\mathcal{A}$ with elements in $\mathcal{B}$. A basis of the product space can be found as the tensor product of the bases of $\mathcal{A}$ and $\mathcal{B}$.

Definition 1 (Product Space). *Given two $\mathbb{F}_q$-subspaces $\mathcal{A}$ and $\mathcal{B}$ of $\mathbb{F}_{q^m}$ with bases $\alpha = (\alpha_1, \dots, \alpha_n)$ and $\beta = (\beta_1, \dots, \beta_m)$, the product space $\langle \mathcal{AB} \rangle$ is the space generated by the set $\{\alpha_i \cdot \beta_j, 1 \leq i \leq n, 1 \leq j \leq m\}$. Its dimension is bounded from above by $n \cdot m$.*

The dimension of the product space is naturally bounded by the product of the dimensions of the two component spaces $\mathcal{A}$ and $\mathcal{B}$. From [4], we also have that the probability that the dimension of the product space is 'full' is given by:

$$\Pr(\dim(\langle \mathcal{AB} \rangle) = n \cdot m) \geq 1 - n\frac{q^{nm}}{q^m}$$

We also recall the concept of *rank multiplication* as described by Loidreau [7]:

Definition 2 (Rank Multipliers). *ß Let $\mathcal{V} = \langle \nu_1, \dots, \nu_\lambda \rangle$ be a λ-dimensional subspace of $\mathbb{F}_{q^m}$. Any invertible matrix $\mathbf{P} \in \mathbb{F}_{q^m}^{nxn}$ with coefficients in $\mathcal{V}$ is a rank multiplier of weight λ with the property: $|\mathbf{xP}|_r \leq \lambda \cdot |\mathbf{x}|_r, \quad \forall \mathbf{x} \in \mathbb{F}_{q^m}^n$. The inverse of a rank multiplier $\mathbf{P}^{-1} \in \mathbb{F}_{q^m}^{nxn}$, on the other hand, has elements that are no longer confined to the subspace $\mathcal{V}$, but rather belong to all of $\mathbb{F}_{q^m}$.*

This is a direct consequence of the properties of a product space since the components of any $\mathbf{xP}$ belong to the product space $\langle \mathcal{XV} \rangle$, where $\mathcal{X}$ is the support of $\mathbf{x}$ and has dimension at most $|\mathcal{X}|_r\lambda$.

We can quantify the probability that the rank of $\mathbf{xP}$ is exactly $\lambda|\mathbf{x}|_r$ using the result obtained for product spaces in [1]. If $\mathbf{x}$ is of rank r (i.e., the support of $\mathbf{x}$ has r linearly independent elements) and the support $\mathcal{V}$ of $\mathbf{P}$ has λ independent elements, the probability that the product space has dimension exactly λr (and thus $|\mathbf{xP}|_r = r\lambda$) is given by

$$\Pr(\dim(\langle \mathcal{XV} \rangle) = r\lambda) \geq 1 - \lambda q^{\lambda r - m}, \tag{1}$$

which is exponentially close to 1 for the large values of q required for RankSign. The rank amplifier $\mathbf{P}$ can equivalently be viewed as the parity-check matrix of a rate-0 LRPC code with weight λ, i.e., a matrix whose entries all belong to a small λ-dimensional subspace of $\mathbb{F}_{q^m}$.

2.2 Bounds in the Rank Metric

Let $S(n, m, q, i)$ denote the cardinality of a sphere of radius i, i.e., the number of words of $\mathbb{F}_{q^m}^n$ with weight exactly i. In the rank metric case, this corresponds to the number of $m \times n$ matrices over $\mathbb{F}_q$ that have rank i, which is

$$S(n, m, q, i) = \prod_{j=0}^{i-1} \frac{(q^m - q^j)(q^n - q^j)}{q^i - q^j}.$$

By definition, the volume of a ball of radius r is given by $V(n, m, q, r) = \sum_{i=0}^{r} S(n, m, q, i)$, i.e., the number of words with weight up to (and including) r.

The Rank-Gilbert Varshamov (RGV) bound can be defined similarly to the Hamming case as follows.

Definition 3 (Rank Gilbert-Varshamov Bound). *Let C be an $[n,k]_{q^m}$ linear rank metric code of length n and dimension k over the extension field $\mathbb{F}_{q^m}$. The rank Gilbert-Varshamov bound $RGV(n,k,m,q)$ for C is the smallest integer r such that $V(n,m,q,r) > q^{(n-k)m}$, i.e., the smallest r for which the number of words with weight up to r exceeds the number of syndromes. In the general case ($m \neq n$):*

$$RGV(n,m,k) \sim \frac{m+n-\sqrt{(m-n)^2+4km}}{2}.$$

Finally, the Singleton bound can be transferred to the rank metric as in [3]:

Definition 4 (Rank Singleton Bound). *The minimum rank distance d_R of an $[n,k]_{q^m}$ linear rank metric code is bounded by:*

$$d_R \leq n-k+1.$$

If $n > m$, a tighter bound can be obtained as:

$$d_R \leq \frac{m}{n}(n-k)+1.$$

2.3 A Difficult Problem in the Rank Metric

To define a cryptographic scheme, we need a problem that is in general hard to solve but which becomes feasible given some secret trapdoor information. One such problem is the rank-syndrome decoding problem, which was recently proven hard by a probabilistic reduction in [6].

Definition 5 (Rank-Syndrome Decoding Problem). *Given an $(n-k)\times n$ matrix $\mathbf{H}$ over $\mathbb{F}_{q^m}$, $\mathbf{s} \in \mathbb{F}_{q^m}^{n-k}$ and an integer r, the RSD problem consists of finding an error $\mathbf{e} \in \mathbb{F}_{q^m}^n$ such that $\mathbf{H}\mathbf{e}^T = \mathbf{s}$ and $|\mathbf{e}| = r$.*

Note that if $r = |\mathbf{e}|$ is chosen to be the RGV bound, we expect the system $\mathbf{H}\mathbf{e}^T = \mathbf{s}$ to have on average one solution. The security of RankSign relies on the masking of the low-rank parity check matrix $\mathbf{H}$ since knowledge of this low-dimensional subspace is what makes decoding efficient.

3 LRPC Codes

LRPC codes were established in [4] as the rank-metric-analogue to classical LDPC codes of the Hamming metric. They are well-suited for cryptographic applications due to their weak algebraic structure.

Definition 6 (LRPC Code). *A Low-Rank Parity Check code of rank d, length n and dimension k over $\mathbb{F}_{q^m}$ is a code that has as parity check matrix an $(n-k)\times n$ matrix $\mathbf{H}$ such that the subspace $\mathcal{F} \subseteq \mathbb{F}_{q^m}$ generated by its coefficients h_{ij} has dimension at most d.*

Put differently, the elements of $\mathbf{H}$ are not drawn from the whole of $\mathbb{F}_{q^m}$, but rather a small d-dimensional subspace that we denote as $\mathcal{F}$.

Efficient decoding of LRPC codes is at the heart of the RankSign signature scheme. In order to be able to decode above the RGV bound, the classical LRPC decoding algorithm is adapted to an *erasure decoder*, which can not only correct rank errors but also rank erasures.

3.1 Erasure Decoder for LRPC Codes

In the decoding of LRPC codes, given a parity check matrix $\mathbf{H}$ and a syndrome $\mathbf{s}$, we are asked to find an error $\mathbf{e}$ of small rank weight r such that $\mathbf{He}^T = \mathbf{s}$. In *erasure decoding*, some additional side information is incorporated, e.g., on the subspace spanned by the elements of the error vector. In the case of RankSign, we define a randomly chosen t-dimensional *erasure space* $\mathcal{T}$ and impose the condition $\mathcal{T} \subset \mathcal{E}$, where $\mathcal{E}$ space generated by the elements of $\mathbf{e}$.

Once the erasure space $\mathcal{T}$ is fixed, we can define the set of $\mathcal{T}$-decodable syndromes, which are the syndromes that are efficiently decodable to a low-weight error whose support contains $\mathcal{T}$:

Definition 7 ($\mathcal{T}$-Decodability). *Let $\mathbf{F}_1$ and $\mathbf{F}_2$ be two linearly independent elements of $\mathcal{F}$ and let $\mathcal{T}$ be a fixed erasure space of small dimension t. A syndrome $\mathbf{s} \in \mathbb{F}_{q^m}^{n-k}$ is $\mathcal{T}$-decodable if there exists a subspace $\mathcal{E}$ of $\mathbb{F}_{q^m}$ with dimension r that fulfills the following conditions:*

1. $\dim(\langle \mathcal{F}\mathcal{E} \rangle) = \dim(\mathcal{F}) \dim(\mathcal{E})$
2. $\dim(\mathbf{F}_1^{-1}\langle \mathcal{F}\mathcal{E} \rangle \cap \mathbf{F}_2^{-1}\langle \mathcal{F}\mathcal{E} \rangle) = \dim(\mathcal{E})$
3. $supp(\mathbf{s}) \subset \langle \mathcal{F}\mathcal{E} \rangle \wedge supp(\mathbf{s}) + \langle \mathcal{F}\mathcal{T} \rangle = \langle \mathcal{F}\mathcal{E} \rangle$

The decoding algorithm can be divided into two phases: First, the error space $\mathcal{E}$ is recovered under the condition $\mathcal{T} \subset \mathcal{E}$. Second, the system $\mathbf{He}^T = \mathbf{s}$ is rewritten using the basis of the product space $\langle \mathcal{F}\mathcal{E} \rangle$ using the error space estimated in the previous step, which gives just enough equations to solve for $\mathbf{e}$. Decoding is performed under the assumption that $\mathbf{s}$ is $\mathcal{T}$-decodable according to Definition 7. If s happens to be non-$\mathcal{T}$-decodable, the decoder reports a failure and can be queried again with a different randomly chosen subspace $\mathcal{T}$. However, it can be shown that the parameters can be chosen such that any syndrome is $\mathcal{T}$-decodable with probability arbitrarily close to 1.

Recovering the Error Space. The decoder first computes the product space $\mathcal{B} = \langle \mathcal{F}\mathcal{T} \rangle$ and the subspace $\mathcal{S} = \langle \mathcal{B} \cup \{s_1, \ldots, s_{n-k}\} \rangle$. According to condition (3) of $\mathcal{T}$-decodability the coordinates s_i of s all belong to the space $\langle \mathcal{F}\mathcal{E} \rangle$and $\mathcal{S}$ is equal to $\langle \mathcal{F}\mathcal{E} \rangle$ for some $\mathcal{E}$. Now, from the definition of product spaces, we have that $\mathcal{E} \subseteq \mathbf{F_1}^{-1}\langle \mathcal{F}\mathcal{E} \rangle \cap \mathbf{F_2}^{-1}\langle \mathcal{F}\mathcal{E} \rangle$ and using condition (2) of $\mathcal{T}$-decodability together with $\mathcal{S} = \langle \mathcal{F}\mathcal{E} \rangle$ we know that in fact $\mathcal{E} = \mathbf{F_1}^{-1}\mathcal{S} \cap \mathbf{F_2}^{-1}\mathcal{S}$.

Recovering the Error. Now that we have obtained $\mathcal{E}$, we can solve the system $\mathbf{He}^T = \mathbf{s}$ with the additional constraint that $\mathbf{e} \in \mathcal{E}^n$. First, we rewrite $\mathbf{He}^T = \mathbf{s}$ as a linear system over the small field $\mathbb{F}_q$ by expanding the elements of $\mathbf{H}$,$\mathbf{e}$ and $\mathbf{s}$ (which are from $\mathbb{F}_{q^m}$) to vectors over $\mathbb{F}_q$ according to the bases $\mathcal{F}$, $\mathcal{E}$ and $\mathcal{S} = \langle \mathcal{FE} \rangle$, respectively. Decoding can also be sped up by precomputing the decoding matrix $\mathbf{D}$, which will be introduced in Sect. 4.1.

Algorithm 1. Erasure decoder for LRPC codes

Input: A $\mathcal{T}$-decodable syndrome $\mathbf{s}' \in \mathbb{F}_{q^m}^{n-k}$, decoding matrix $\mathbf{D}$, erasure space $\mathcal{T}$, $\mathcal{F} = \langle h_{ij} \rangle$
Output: $\mathbf{e}'$ of weight r such that $\mathbf{He}'^T = \mathbf{s}'$
1: $\mathcal{B} \leftarrow (\mathbf{F}_i \mathbf{T}_j)$, a basis of $\langle \mathcal{FT} \rangle$
2: $\mathcal{S} \leftarrow \langle \mathcal{B} \cup \{s_1, \ldots, s_{n-k}\} \rangle$
3: $\mathcal{E} \leftarrow \mathbf{F_1}^{-1} \mathcal{S} \cap \mathbf{F_2}^{-1} \mathcal{S}$
4: Express $\mathbf{s}'$ in the basis $(\mathbf{F}_i \mathbf{E}_j)$ of $\langle \mathcal{FE} \rangle$
5: Express the coordinates of $\mathbf{e}'$ in a basis $\mathbf{e}'_i = \sum_{j=1}^{r} \mathbf{e}''_{ij} \mathbf{E}_j$
6: Solve $\mathbf{He}'^T = \mathbf{s}'$ with nr unknowns (the $\mathbf{e}''_{ij}$) and $(n-k)rd$ equations.
7: **return** $\mathbf{e}'$

As long as the syndrome s' is $\mathcal{T}$-decodable, the algorithm will always return a valid $\mathbf{e}'$. When choosing a random syndrome to decode, there is some probability that one of the conditions of $\mathcal{T}$-decodability is not fulfilled, causing the algorithm to fail. In the RankSign scheme, parameters were chosen such that this case is overwhelmingly unlikely. The full erasure decoder is illustrated in Algorithm 1.

4 The RankSign Signature Scheme

RankSign is a hash-and-sign signature scheme based on the decoding of LRPC codes. A signature is generated by interpreting the message (or rather a fixed-length hash thereof) as the syndrome of a public LRPC code. The signer can decode the message to a low-weight error using the private key, which contains knowledge about the structure of the LRPC code, i.e., the low-dimensional subspace from which the elements of its parity-check matrix are drawn. A typical concern in hash-and-sign schemes is the density of decodable syndromes, meaning that not every hashed message corresponds to a syndrome that can be efficiently decoded. In fact, if the required weight of the error vector is chosen to be small, say on the RGV bound, we expect this instance of the syndrome decoding problem to have only one solution on average, making the decoding infeasible using the 'standard' error decoder. By using the erasure decoder introduced in Sect. 3.1, it is possible to instead decode to an error vector whose weight lies above the RGV bound while imposing the additional condition that the support of this error contains a randomly chosen low-dimensional 'erasure space' $\mathcal{T}$.

4.1 KeyGen

During the decoding step, an error $\mathbf{e}'$ of small weight r will be computed such that $\mathbf{He}'^T = \mathbf{s}'$. Recall that the elements of $\mathbf{H}$ are drawn from the d-dimensional space $\mathcal{F}$ and the error $\mathbf{e}'$ has a basis $\mathcal{E}$ with dimension r. The elements of the syndrome $\mathbf{s}'$ can then be written as elements of the product space $\langle \mathcal{F}\mathcal{E} \rangle$, leading to a system $\mathbf{Ae}'^T = \mathbf{s}'$ with nr unknowns (the elements of $\mathbf{e}'$ in the basis $\mathcal{E}$) and $(n-k)rd$ equations, where the matrix $\mathbf{A}$ is obtained by representing each element of $\mathbf{H}$ in the basis of the product space $\langle \mathcal{F}\mathcal{E} \rangle$. From $\mathbf{A}$, we can derive the decoding matrix $\mathbf{D} = \mathbf{A}^{-1}$ and store it in the secret key. Note that this reduces the equation-solving part of the decoder to a simple multiplication. Since $\mathbf{A}$ solely depends on $\mathbf{H}$ and the weight r, we can do this precomputation in the key generation step. Moreover, in order to ensure that $\mathbf{A}$ exists and is invertible, we can flip the problem: Generate $\mathbf{A}$ uniformly at random and derive $\mathbf{H}$ from it.

$\mathbf{H}$ and $\mathbf{A}$ are related by the following equation, see [1]:

$$h_{ijv} = a_{u+(v-1)r+(i-1)rd,\; u+(j-1)r}, \tag{2}$$

where $\mathbf{A} = (h_{ijv})$ is the matrix $\mathbf{H}$ unfolded in the basis $\mathcal{F}$ of dimension d and $\mathbf{A} = (a_{ij})$ with $i \in \{1, \ldots, n-k\}$, $j \in \{1, \ldots, n\}$.

The elements of the invertible masking matrix $\mathbf{Q}$ are drawn uniformly at random from $\mathbb{F}_{q^m}$, while the right scrambler $\mathbf{P}$ has only elements from the base field $\mathbb{F}_{q^m}$. Additionally, t random columns $\mathbf{R}$ over $\mathbb{F}_{q^m}$ are appended to the parity-check matrix $\mathbf{H}$ to increase the distance of the dual code (which is not affected by the multiplication with $\mathbf{P}$ from the base field). The key generation is summarized in Algorithm 2.

Algorithm 2. Key generation

Output: pk = $[\mathbf{H}_{pub}, \mathrm{G}]$, sk = $[\mathbf{F}, \mathbf{Q}, \mathbf{P}, \mathbf{D}]$
1: Choose $\mathcal{F} = \langle \mathbf{F}_1, \ldots, \mathbf{F}_d \rangle$, a random d-dimensional subspace of $\mathbb{F}_{q^m}$
2: Choose $\mathbf{H}$ of size n x n invertible in $\mathbb{F}_{q^m}$ with coefficients in $\mathbb{F}_q$.
3: $\mathbf{D} \leftarrow \mathbf{A}^{-1}$
4: Compute $\mathbf{H}$ from $\mathbf{A}$ according to Eq. 2
5: $\mathbf{P} \xleftarrow{\$} GL_n(\mathbb{F}_q)$
6: $\mathbf{R} \xleftarrow{\$} \mathbb{F}_{q^m}^{(n-k)\times t}$
7: $\mathbf{H}_{pub}$ = ECHELONIZE$(\mathbf{HP}^{-1})$ ▷ negligible failure probability $\sim q^{-m}$
8: $\mathbf{Q} \leftarrow GL_{n-k}(\mathbb{F}_{q^m})$ such that $\mathbf{Q}(\mathbf{R}|\mathbf{H})\mathbf{P}^{-1} = \mathbf{H}_{pub}$
9: sk $\leftarrow [\mathbf{H}_{pub}, \mathrm{G}]$
10: pk $\leftarrow [\mathbf{F}, \mathbf{Q}, \mathbf{P}, \mathbf{D}]$
11: **return** sk, pk

4.2 Sign

To authenticate a message, the signer first chooses a random seed of length l and computes s, the hash of the message with respect to the seed. Recall that the public key is constructed as $\mathbf{H}_{pub} = \mathbf{Q}(\mathbf{H}|\mathbf{R})\mathbf{P}^{-1}$, hence $\mathbf{Q}(\mathbf{R}|\mathbf{H})\mathbf{P}^{-1}\mathbf{e}^T = \mathbf{s}$. The final error vector **e** must be assembled from two parts: One of length $n - t$ that is decoded with respect to the private parity check matrix H, and one of length t, that accounts for the random columns **R** appended to **H**. From [4] we know that the modified syndrome s' will be decodable with respect to a randomly chosen t-dimensional subspace $\mathcal{T}$ with high probability. Should the syndrome ever be non-$\mathcal{T}$-decodable, the decoder will be queried again with a new randomly chosen subspace $\mathcal{T}$. The result of the decoder is a vector $\mathbf{e}'$ of rank weight r, whose weight is not affected by the multiplication with **P**, since multiplication with **P** is an isometry for the rank metric. A more detailed description of the signature generation is found in Algorithm 3.

Algorithm 3. Signature of a message

Input: message m, pk = $[\mathbf{H}_{pub}, \text{G}]$, sk = $[\mathbf{F}, \mathbf{Q}, \mathbf{P}, \mathbf{D}]$
Output: seed, signature **e** such that $\mathbf{H}_{pub}\,\mathbf{e}^T = \mathbf{s}$ and $|\mathbf{e}|_r = r$

1: seed $\xleftarrow{\$} \{0,1\}^l$
2: s $\leftarrow$ G(m, seed)
3: Choose $\mathbf{e}' \in \mathbb{F}_{q^m}^{n+t}$ of weight t' uniformly at random.
4: Choose $\mathcal{T} = \langle T_1, \ldots, T_t \rangle$, a random t-dimensional subspace of $\mathbb{F}_{q^m}$
5: $\mathbf{s}' \leftarrow \mathbf{s} - \mathbf{e}'\mathbf{H}_{pub}{}^T$
6: $\mathbf{s}'' \leftarrow \mathbf{s}'(\mathbf{Q}^T)^{-1} - (T_t, \ldots T_t)\mathbf{R}^T$
7: $\mathbf{e}'' \leftarrow \text{Decode}(\mathbf{s}'', \mathbf{D}, \mathcal{T}, \mathcal{F})$ ▷ If $\mathbf{s}''$ is not $\mathcal{T}$-decodable, return to 1
8: $\mathbf{e} \leftarrow \mathbf{e}' + (T_1, \ldots, T_t|\mathbf{e}'')(\mathbf{P}^T)^{-1}$
9: **return** e, seed

4.3 Verify

Given the public key pk = $[\mathbf{H}_{pub}, \text{G}]$, the document m, and its associated signature $[\mathbf{e}|seed]$, the verifier checks that:

- $\mathbf{H}_{pub}\,\mathbf{e}^T = \text{G(m,seed)}$
- $|\mathbf{e}|_r \leq \lambda r$.

5 Attack by Debris-Alazard and Tillich

Shortly after its submission to the NIST competition, the RankSign scheme was broken by an attack by Debris-Alazard and Tillich [2] in which it was shown that the public code contains many codewords of rank weight 2 that could be efficiently recovered using Gröbner basis techniques.
First, we recall a general result given in the paper on the minimum distance of *any* LRPC code.

Theorem 1 (Low-rank codewords in LRPC codes). *Let $\mathcal{C}$ be an $[n,k]$ LRPC code over $\mathbb{F}_{q^m}$ defined by a parity-check matrix* $\mathbf{H}$ *whose entries lie in a subspace $\mathcal{F}$ of $\mathbb{F}_{q^m}$. If there exists another subspace $\mathcal{F}' \subseteq \mathbb{F}_{q^m}$ with the property*

$$(n-k)\dim(\langle \mathcal{F}\mathcal{F}' \rangle) < n \dim(\mathcal{F}'),$$

then $\mathcal{C}$ contains non-zero codewords whose support is included in $\mathcal{F}'$ (i.e., their weight is at most $\dim(\mathcal{F}')$*).*

This theorem can be derived from observing the following: Any codeword in $\mathcal{C}$ satisfies $\mathbf{H}\mathbf{c}^T = 0$. Supposing that $\mathbf{c}$ has its support in $\mathcal{F}'$, it can be expressed in a basis of $\mathcal{F}'$. $\mathbf{H}c^T = 0$ gives $n-k$ equations which can be expressed in a basis of $\langle \mathcal{F}\mathcal{F}' \rangle$, hence we obtain $(n-k)\dim(\langle \mathcal{F}\mathcal{F}' \rangle)$ equations over $\mathbb{F}_q$ with $n \cdot \dim(\mathcal{F}')$ unknowns. A solution exists (i.e. $\mathcal{C}$ contains codewords whose support is in $\mathcal{F}'$) if $n \dim(\mathcal{F}') > (n-k)\dim(\langle \mathcal{F}\mathcal{F}' \rangle)$.

For the following, let $\mathcal{F}'$ be a subspace of $\mathcal{F}$ with dimension 2. If $\{x_1, x_2, \ldots, x_d\}$ is a basis for $\mathcal{F}$, let $\{x_1, x_2\}$ be the basis of $\mathcal{F}'$. Now observe that the dimension of the product space $\langle \mathcal{F}\mathcal{F}' \rangle$ can be at most $\dim(\mathcal{F}') \cdot \dim(\mathcal{F}) - 1 = 2d-1$ using the definition of product spaces and the fact that $x_1 x_2 = x_2 x_1$. Theorem 1 gives the result that $\mathcal{C}$ contains codewords of weight $\dim(\mathcal{F}')$ if $n\dim(\mathcal{F}') > (n-k)\dim(\langle \mathcal{F}\mathcal{F}' \rangle)$. Together with $\dim(\langle \mathcal{F}\,\mathcal{F}' \rangle) \leq 2d-1$ we get that $\mathcal{C}$ contains codewords of weight 2 if $2n > (n-k) \cdot (2d-1)$.

Denote by $\mathcal{C}$ the code defined by $\mathbf{H}$ and let

$$\mathcal{C}' = \{\mathbf{c} \in \mathcal{C} : c_i \in \mathcal{F}', 1 \leq i \leq n+t\}$$

i.e., the subset of all codewords in $\mathcal{C}$ whose support is in $\mathcal{F}'$ and that consequently have weight at most 2. These form an $\mathbb{F}_q$-subspace of $\mathbb{F}_{q^m}$ that has dimension $\geq 2n - (n-k) \cdot (2d-1)$.

Now, for the original RankSign scheme, we have the constraint on the parameters that $n = (n-k)d$, and thus, for this special case

$$\dim_{\mathbb{F}_q}(\mathcal{C}') \geq \frac{n}{d}.$$

Since clearly $\mathcal{C}'$ in $\mathcal{C}$, this means that there exists a large amount of weight-2 codewords in the secret code $\mathcal{C}$. Finally, it was shown that also $\mathcal{C}'$ in $\mathcal{C}'_{pub}$, i.e., the low-weight codewords from the secret code carry over to the public code.

Recall the public parity-check matrix $\mathbf{H}_{pub}$ in RankSign is constructed as follows:

$$\mathbf{H}_{pub} = \mathbf{Q}[\mathbf{R}|\mathbf{H}]\mathbf{P}$$

where $\mathbf{Q}$ is an invertible matrix over $\mathbb{F}_{q^m}$, $\mathbf{R}$ is a matrix over $\mathbb{F}_{q^m}$ adding t random columns to $\mathbf{H}$ and $\mathbf{P}$ is an invertible matrix over the base field $\mathbb{F}_q$. Let $\mathcal{C}_{pub}$ be the public code generated by $\mathbf{H}_{pub}$.

Define now

$$\mathcal{C}'_{pub} = \{(0_t, \mathbf{c})(P^{-1})^T : \mathbf{c} \in \mathcal{C}'\}.$$

Clearly the dimension of $\mathcal{C}'_{pub}$ is the same as $\mathcal{C}'$ and its codewords $\mathbf{c}'$ have their entries in $\mathcal{F}'$ since $\mathbf{P}$ is only a permutation matrix over $\mathbb{F}_q$. Finally, it can be shown that $\mathcal{C}'$ in $\mathcal{C}'_{pub}$ and thus the public code contains codewords of rank 2:

$$\begin{aligned}\mathbf{H}_{pub}\mathbf{c}'^T &= \mathbf{Q}[\mathbf{R}|\mathbf{H}]\mathbf{P}\mathbf{P}^{-1}\\ &= \mathbf{Q}[\mathbf{R}|\mathbf{H}]\mathbf{c}^T\\ &= 0 \quad \text{(since } \mathbf{c} \text{ belongs to } \mathcal{C} \text{ defined by } \mathbf{H}).\end{aligned}$$

Having identified that the public code contains many codewords of low-rank weight (weight 2 in the case of the RankSign parameters), these codewords could be recovered using algebraic techniques and enable an attacker to sign like a legitimate user.

In summary, the proposed attack hinges on the fact that the LRPC code defined by the low-rank parity check matrix $\mathbf{H}$ contains many low-weight codewords that cannot be masked by the scrambling matrix $\mathbf{P}$ since its entries are drawn from the base field $\mathbb{F}_q$.

6 Repairing RankSign

In the following, we propose a repair to the RankSign scheme by replacing $\mathbf{P}$ with the inverse of a rank-multiplier of weight λ. This can equivalently be described as multiplying the generator matrix of the public code by a rank multiplier, thus vastly reducing the number of low-weight codewords according to Theorem 1.

6.1 KeyGen

The key generation remains very similar to what was discussed in Algorithm 2. The two notable changes are the replacement of $\mathbf{P} \in \mathbb{F}_q^{n\times n}$ in step 5 by $\mathbf{P}^{-1} \in \mathbb{F}_{q^m}^{n\times n}$, the inverse of a rank multiplier of weight λ as shown in Definition 2 and the omission of step 6, since the random columns $\mathbf{R}$ are no longer required due to the masking by $\mathbf{P}^{-1}$, which behaves like a random matrix over $\mathbb{F}_{q^m}$.

6.2 Sign

Algorithm 4 illustrates the modified signature generation.
Since we no longer have to consider the random columns $\mathbf{R}$, the algorithm becomes a little simpler. Note that since $\mathbf{P}$ has been replaced by a rank multiplier, the weight of the error in step 6 will now be increased by a factor of λ.
Correctness. The returned vector $\mathbf{e}$ fulfills the property $\mathbf{H}_{pub}\,\mathbf{e}^T = \mathbf{s}$ since

$$\begin{aligned}\mathbf{H}_{pub}\,\mathbf{e}^T &= \mathbf{s}\\ \mathbf{Q}\mathbf{H}\mathbf{P}^{-1}(\mathbf{P}\,\mathbf{e}')^T &= \mathbf{s}\\ \mathbf{H}\mathbf{e}'^T &= \mathbf{s}',\end{aligned}$$

which is true by the definition of the erasure decoder. We also have $|\mathbf{e}|_r = \lambda r$ with high probability since $\mathbf{e}'$ has rank weight r by definition and most likely $|\mathbf{P}\,\mathbf{e}'|_r = \lambda r$ from (1).

Algorithm 4. Signature of a message

Input: message m, pk = $[\mathbf{H}_{pub}, \mathrm{G}]$, sk = $[\mathbf{F}, \mathbf{Q}, \mathbf{P}, \mathbf{D}]$
Output: seed, signature $\mathbf{e}$ such that $\mathbf{H}_{pub}\,\mathbf{e}^T = \mathbf{s}$ and $|\mathbf{e}| = \lambda r$

1: seed $\xleftarrow{\$} \{0,1\}^l$
2: $\mathbf{s} \leftarrow \mathrm{G}(\mathrm{m}, \mathrm{seed})$
3: $\mathbf{s}' \leftarrow \mathbf{Q}^{-1}\mathbf{s}$
4: Choose $\mathcal{T} = \langle T_1, \ldots, T_t \rangle$, a random t-dimensional subspace of $\mathbb{F}_{q^m}$
5: $\mathbf{e}' \leftarrow \textsc{Decode}(\mathbf{s}', \mathbf{D}, \mathcal{T}, \mathcal{F})$ ▷ If $\mathbf{s}'$ is not $\mathcal{T}$-decodable, return to 1
6: $\mathbf{e} \leftarrow \mathbf{P}\,\mathbf{e}'$
7: **return** $\mathbf{e}$, seed

6.3 Verify

The verification algorithm remains unchanged, except for the fact that the verifier now has to check whether $|\mathbf{e}|_r \leq \lambda r$.

6.4 Key and Signature Size

To represent one element in $\mathbb{F}_{q^m}$ we need $\log_2(q) \cdot m$ bits of storage. Since $\mathbf{H}_{pub} = [I_{n-k}|\mathbf{R}']$ is in systematic form, it is sufficient to store $\mathbf{R}'$ of size $(n-k)\cdot k$, resulting in a signature size of

$$\mathrm{size}(\mathbf{H}_{pub}) = (n-k)\cdot km\log_2(q) \text{ bits.}$$

The signature e can be stored in one of two ways, depending on the size of the parameter λ. Either as n elements of $\mathbb{F}_{q^m}$, making the signature size $\log_2(q)\cdot mn$, or as n linear combinations of supp($\mathbf{e}$). We can obtain the support of $\mathbf{e}$ from the product space $\langle \mathcal{V}\mathcal{E} \rangle$, which has dimension at most $\lambda \cdot r$. To save additional space, the support of $\mathbf{e}$ can be represented in row echelon form, giving a support size of $\log_2(q)((m-\lambda r)m)$ bits. In summary, the size of the signature is given by:

$$\mathrm{size}(\mathbf{e}) = \min\{\log_2(q)\cdot mn,\ \log_2(q)\cdot(\lambda rn + (m-\lambda r)m)\} \text{ bits.}$$

Thus, the size of the signature increases by a factor of λ while the key size remains the same (for the same choice of n, m, k).

6.5 Discussion of Parameters

The original RankSign scheme imposes the following strict conditions on its parameters:

- $m = (r-t)(d+1)$
- $(n-k) = d(r-t-t')$
- $n = (n-k)d$,

where r is the weight of the decoded error, t is the dimension of the erasure space $\mathcal{T}$ (which is equal to the number of random columns in $\mathbf{R}$) and t' is the weight of $\mathbf{e}'$ in the signing Algorithm 3. These parameters ensure that generic decoding of the public code is hard (it is required to stay above the Singleton bound at least) and that the density of decodable syndromes remains high. For our adapted version, we now have the weight of the error as $r' = \lambda r$, which leads to an increase of m and n to remain in the range of feasible parameters. In turn, the weight of the error $\mathbf{e}$ must increase to remain above the RGV bound and ensure a high decodable density.

To further illustrate this point, consider the *public* decoding problem, $\mathbf{H}_{pub}\mathbf{e}^T = \mathbf{s}$ where $\mathbf{e}$ must be of weight $\leq \lambda r$. To remain resilient to generic decoding, it is necessary to impose $\lambda r \leq \frac{m(n-k)}{n}$ (Singleton bound) to avoid even a basic polynomial-time attack. The public decoding problem thus gives constraints on m, n, k, and r. Conversely, in the *private* decoding problem $\mathbf{H}\mathbf{e}'^T = \mathbf{s}'$, $\mathbf{s}'$ must be decoded to an error $\mathbf{e}'$ of weight r. This parameter r must be chosen above (and optimally near) the RGV bound to ensure any syndrome's decodability. Hence, the *private* decoding problem gives another constraint on those same parameters n, m, k but forces them in the other direction. This leads to a sort of avalanche effect, where the parameters grow until the distance between the Singleton bound and the RGV bound becomes large enough, leading to infeasible parameter sizes. It remains an open problem how the scheme can be further adapted to retain a reasonably-sized set of parameters.

Finally, we conclude with two open questions that might lead to a feasible set of parameters for the proposed scheme:

1. Is it always possible to find a 'complementary' erasure space $\mathcal{V}$, such that the product space $\langle \mathcal{T}\mathcal{V} \rangle$ has the same dimension as $\mathcal{T}$? Put more generally, under which conditions does such a 'complementary' subspace exist for any given subspace of $\mathbb{F}_{q^m}$, and how does it affect the performance of the decoder?
2. Is it possible to define a *partial* rank multiplier? If so, which effects does it have on a given code and its dual? Is the inverse of this partial rank amplifier distinguishable from a random matrix over $\mathbb{F}_{q^m}$, and if so, does this weaken its masking property?

7 Conclusion

We considered the use of rank multipliers to repair the digital signature scheme RankSign. The initial proposal of RankSign was broken by an attack targeting low-weight codewords in the public code that were insufficiently masked in the key generation step. The distance of the public code can be increased by replacing the scrambling matrix over the small field $\mathbb{F}_q$ with the inverse of a rank multiplier. Additionally, the public parity-check matrix is now masked by a matrix that behaves like a random matrix over $\mathbb{F}_{q^m}$, making the addition of random columns to increase the distance of the dual code obsolete. The selection of parameters, however, poses a challenge. The rank multiplier causes a large increase in the weight of the error that is decoded with respect to the public parity-check matrix, which leads to an increase in the overall parameter size to be resistant to generic decoding attacks.

Finally, we pose two open questions that could lead to feasible parameters for the repaired scheme: First, is it possible to construct the rank multiplier **P** in such a way that the weight of the error is amplified by a smaller amount? This could be achieved by choosing the erasure space $\mathcal{T}$ complementary to the support of **P**, such that only part of the error is amplified, though it is unclear how this would affect the success probability of the erasure decoder. Second, is it possible to replace **P** with a *partial* rank amplifier? This would probably involve splitting the rank multiplier into a block whose elements are from $\mathbb{F}_{q^m}$ and a block whose elements are drawn from the small field $\mathbb{F}_q$, though it is unclear how this would affect the masking of both the public and dual of the public code.

Acknowledgements. This work was supported by the German Research Foundation (Deutsche Forschungsgemeinschaft, DFG) under Grant No. WA3907/4-1 and Munich Aerospace e.V. under the research grant "Multiaccess and Security Coding for Massive IoT Satellite Systems".

References

1. Aragon, N., Gaborit, P., Hauteville, A., Ruatta, O., Zémor, G.: Low rank parity check codes: new decoding algorithms and applications to cryptography. IEEE Trans. Inf. Theory **65**(12), 7697–7717 (2019)
2. Debris-Alazard, T., Tillich, J.-P.: Two attacks on rank metric code-based schemes: RankSign and an IBE scheme. In: Peyrin, T., Galbraith, S. (eds.) ASIACRYPT 2018. LNCS, vol. 11272, pp. 62–92. Springer, Cham (2018). https://doi.org/10.1007/978-3-030-03326-2_3
3. Gabidulin, E.M.: Theory of codes with maximum rank distance. Problemy peredachi informatsii **21**(1), 3–16 (1985)
4. Gaborit, P., Murat, G., Ruatta, O., Zémor, G.: Low rank parity check codes and their application to cryptography. In: Proceedings of the Workshop on Coding and Cryptography WCC, vol. 2013 (2013)
5. Gaborit, P., Ruatta, O., Schrek, J., Zémor, G.: RankSign: an efficient signature algorithm based on the rank metric. IACR Cryptology ePrint Archive, p. 766 (2013)
6. Gaborit, P., Zémor, G.: On the hardness of the decoding and the minimum distance problems for rank codes. IEEE Trans. Inf. Theory **62**(12), 7245–7252 (2016)
7. Loidreau, P.: An evolution of GPT cryptosystem. In: International Workshop on Algebraic and Combinatorial Coding Theory (ACCT) (2016)

Fast Gao-Like Decoding of Horizontally Interleaved Linearized Reed–Solomon Codes

Felicitas Hörmann[1,2](✉) and Hannes Bartz[1]

[1] Institute of Communications and Navigation, German Aerospace Center (DLR), Oberpfaffenhofen–Wessling, Germany
{felicitas.hoermann,hannes.bartz}@dlr.de

[2] School of Computer Science, University of St. Gallen, St. Gallen, Switzerland

Abstract. Both horizontal interleaving as well as the sum-rank metric are currently attractive topics in the field of code-based cryptography, as they could mitigate the problem of large key sizes. In contrast to vertical interleaving, where codewords are stacked vertically, each codeword of a horizontally s-interleaved code is the horizontal concatenation of s codewords of s component codes. In the case of horizontally interleaved linearized Reed–Solomon (HILRS) codes, these component codes are chosen to be linearized Reed–Solomon (LRS) codes.

We provide a Gao-like decoder for HILRS codes that is inspired by the respective works for non-interleaved Reed–Solomon and Gabidulin codes. By applying techniques from the theory of minimal approximant bases, we achieve a complexity of $\tilde{\mathcal{O}}(s^{2.373}n^{1.635})$ operations in $\mathbb{F}_{q^m}$, where $\tilde{\mathcal{O}}(\cdot)$ neglects logarithmic factors, s is the interleaving order and n denotes the length of the component codes. For reasonably small interleaving order $s \ll n$, this is subquadratic in the component-code length n and improves over the only known syndrome-based decoder for HILRS codes with quadratic complexity. Moreover, it closes the performance gap to vertically interleaved LRS codes for which a decoder of complexity $\tilde{\mathcal{O}}(s^{2.373}n^{1.635})$ is already known.

We can decode beyond the unique-decoding radius and handle errors of sum-rank weight up to $\frac{s}{s+1}(n-k)$ for component-code dimension k. We also give an upper bound on the failure probability in the zero-derivation setting and validate its tightness via Monte Carlo simulations.

Keywords: Gao-like Decoding · Horizontal Interleaving · Linearized Reed–Solomon Codes · Sum-Rank Metric · Code-Based Cryptography · Minimal Approximant Bases

---

F. Hörmann and H. Bartz acknowledge the financial support by the Federal Ministry of Education and Research of Germany in the programme of "Souverän. Digital. Vernetzt." Joint project 6G-RIC, project identification number: 16KISK022.

A. Esser and P. Santini (Eds.): CBCrypto 2023, LNCS 14311, pp. 14–34, 2023.
https://doi.org/10.1007/978-3-031-46495-9_2

1 Introduction

The American National Institute of Standards and Technology (NIST) started a competition for post-quantum cryptography (PQC) in 2016. After three rounds, the lattice-based key-encapsulation mechanism (KEM) CRYSTALS-Kyber [8] was standardized in July 2022 [3]. Moreover, NIST announced a fourth round to which four KEM candidates advanced: BIKE [4], Classic McEliece [14], HQC [1], and SIKE [9]. SIKE is the only candidate based on hard problems in the area of isogenies and was broken by [17] shortly after NIST's round-4 announcement. The remaining three candidates in this round rely on coding-theoretical problems in the Hamming metric.

In his seminal paper [32] in 1978, McEliece proposed the first code-based cryptosystem, which still serves as a blueprint for most of the recent proposals. The McEliece framework essentially resisted the cryptanalytic effort of 45 years. However, it suffers from large key sizes and is thus not usable in many practical applications.

Rank and Sum-Rank Metric. As the syndrome-decoding problem in the rank metric is harder than its Hamming-metric counterpart [7,10], many McEliece-like schemes based on rank-metric codes as e.g. [18,19,27,28] were considered. Unfortunately, most of them were broken by structural attacks. A new approach is to consider the sum-rank metric which covers both the Hamming and the rank metric as special cases. Even though the gain in terms of key size might not be as large as for the rank metric, it is reasonable to hope that rank-metric attacks cannot be adapted to the sum-rank-metric case [21] and the corresponding systems will remain secure.

Interleaved Codes. Another way to reduce the key size is to use codes with higher error-correction capability. An increased error weight will result in higher complexities for generic attacks like [37] and thus require smaller parameter sizes to achieve the same level of security. One well-known code construction to improve the (burst) error-correction capability is interleaving, where each codeword of the s-interleaved code consists of s vertically or horizontally stacked codewords of s component codes, respectively.

Metzner and Kapturowski [33] showed that vertically interleaved Hamming-metric codes can be efficiently decoded with negligible failure probability as soon as their interleaving order s is high compared to the error weight t. This result was generalized to the rank metric [38,40] and recently also to the sum-rank metric [24]. As no knowledge about the code structure is needed for Metzner–Kapturowski-like decoders, this is a direct generic attack on any code-based cryptosystem based on vertically interleaved codes with high interleaving order. Thus, horizontal interleaving appears to be better suited for cryptographic purposes. This is also reflected in recent proposals as for example in the KEM LowMS [6] that is based on horizontally interleaved Gabidulin codes, in the signature scheme Durandal [5] based on the closely related rank-support-learning

(RSL) problem [10], and in the cryptosystem [2] that makes use of horizontally interleaved low-rank parity-check (LRPC) codes [39].

The cryptanalysis of the underlying hard problems ensures reliable security-level estimates. However, also performance improvements for decoding horizontally interleaved codes have a significant impact as they directly speed up decryption and verification within the corresponding cryptosystems and digital signatures.

HILRS Codes Horizontally interleaved linearized Reed–Solomon (HILRS) codes combine the usage of an alternative decoding metric for higher generic-decoding complexity and the interleaving construction for higher error-correction capability. Both approaches promise to reduce the key size in a McEliece-like setup. The component codes of an HILRS code are linearized Reed–Solomon (LRS) codes which were introduced by Martínez-Peñas in 2018 [29]. Up to now, LRS codes are one of the most studied code families in the sum-rank metric. They are evaluation codes with respect to skew polynomials and form the natural generalization of Reed–Solomon (RS) codes in the Hamming metric and Gabidulin codes in the rank metric.

As the performance of code-based cryptosystems strongly depends on the decoding speed for the underlying codes, fast decoders for HILRS codes are crucial. Currently, the only known decoder for HILRS codes is syndrome-based and has a quadratic complexity in the length sn of the interleaved code (ongoing work [23] extending [22]). It can handle a combination of errors, row erasures, and column erasures.

In contrast, vertically interleaved linearized Reed–Solomon (VILRS) codes, which are constructed by vertically stacking s LRS codewords, allow for decoding with lower complexity $\tilde{\mathcal{O}}(s^{\omega}\mathcal{M}(n)) \subseteq \tilde{\mathcal{O}}(s^{2.373}n^{1.635})$ [12,13]. Here, ω and $\mathcal{M}(n)$ denote the matrix-multiplication coefficient and the cost of multiplying two skew polynomials of degree at most n, respectively, and $\tilde{\mathcal{O}}(\cdot)$ neglects logarithmic factors.

Contributions. This paper presents a Gao-like decoder for HILRS codes. It is based on the original Gao decoder for Reed–Solomon codes in the Hamming metric [20] as well as on its known extensions to Gabidulin codes [45,46] and their horizontally interleaved version [36] in the rank metric. We consider probabilistic unique decoding beyond the unique-decoding radius and derive an upper bound on the decoding-failure probability in the zero-derivation case. We achieve a decoding radius of $\frac{s}{s+1}(n-k)$ for the interleaving order s and for n and k denoting the length and the dimension of the component codes, respectively.

We further show how a major speedup can be obtained by using the theory of minimal approximant bases [11]. The fast variant of the Gao-like decoder achieves subquadratic complexity in the length n of the component codes for a fixed interleaving order s. Particularly, we obtain $\tilde{\mathcal{O}}(s^{\omega}\mathcal{M}(n)) \subseteq \tilde{\mathcal{O}}(s^{2.373}n^{1.635})$ and thus close the performance gap with respect to the decoding of VILRS codes.

Our conceptually new approach to solving the Gao-like key equation results in the fastest known decoder for HILRS codes in the sum-rank metric. Moreover, the special case obtained for the rank metric yields the fastest decoder for horizontally interleaved Gabidulin codes in the rank metric, improving on [36,41,42].

Outline. We start the paper in Sect. 2 by giving basic preliminaries on skew polynomials, on HILRS codes in the sum-rank metric, and on the channel model we consider. Then, we present a Gao-like decoder for HILRS codes in Sect. 3 and analyze its decoding radius, complexity, and failure probability. Section 4 deals with a speedup for the shown decoder that is based on the theory of minimal approximant bases. Finally, we summarize the main results of the paper in Sect. 5 and give an outlook on future work.

2 Preliminaries

We denote the finite field of order q by $\mathbb{F}_q$ and refer to its degree-m extension field by $\mathbb{F}_{q^m}$. We often consider vectors $\boldsymbol{x} \in \mathbb{F}_{q^m}^n$ that are divided into blocks. More precisely, we define a *length partition* of $n \in \mathbb{N}^*$ as the vector $\boldsymbol{n} = (n_1, \dots, n_\ell) \in \mathbb{N}^\ell$ with $\sum_{i=1}^{\ell} n_i = n$ and $n_i > 0$ for all $i = 1, \dots, \ell$. We write $\boldsymbol{x} = (\boldsymbol{x}^{(1)} \mid \cdots \mid \boldsymbol{x}^{(\ell)})$, where the blocks $\boldsymbol{x}^{(i)}$ belong to $\mathbb{F}_{q^m}^{n_i}$ for all $i = 1, \dots, \ell$. Similarly, we write $\boldsymbol{X} = (\boldsymbol{X}^{(1)} \mid \cdots \mid \boldsymbol{X}^{(\ell)})$ for a subdivided matrix $\boldsymbol{X} \in \mathbb{F}_{q^m}^{k \times n}$ with $\boldsymbol{X}^{(i)} \in \mathbb{F}_{q^m}^{k \times n_i}$ for all $i = 1, \dots, \ell$. The $\mathbb{F}_{q^m}$-linear row space of $\boldsymbol{X}$ is denoted by $\langle \boldsymbol{X} \rangle_{q^m}$.

Further choose an $\mathbb{F}_{q^m}$-automorphism θ with fixed field $\mathbb{F}_q$. Note that θ is $\mathbb{F}_q$-linear and satisfies both $\theta(a+b) = \theta(a)+\theta(b)$ and $\theta(a \cdot b) = \theta(a) \cdot \theta(b)$ for arbitrary $a, b \in \mathbb{F}_{q^m}$. Moreover, we consider a map $\delta : \mathbb{F}_{q^m} \to \mathbb{F}_{q^m}$ for which the equalities $\delta(a+b) = \delta(a) + \delta(b)$ and $\delta(ab) = \delta(a)b + \theta(a)\delta(b)$ hold for all $a, b \in \mathbb{F}_{q^m}$. In the finite-field setting, all such θ-derivations δ are inner derivations [29], i.e., they have the form $\delta = \gamma(\mathrm{Id} - \theta)$ for a parameter $\gamma \in \mathbb{F}_{q^m}$ and the identity Id.

The automorphism θ and the derivation δ give rise to a partition of $\mathbb{F}_{q^m}$ with respect to (θ, δ)-conjugacy [25]. Namely, two elements $a, b \in \mathbb{F}_{q^m}$ are conjugate if there is a nonzero $c \in \mathbb{F}_{q^m}^*$ with

$$a^c := \theta(c)ac^{-1} + \delta(c)c^{-1}.$$

The conjugacy class of an element $a \in \mathbb{F}_{q^m}$ is denoted by $\mathcal{C}(a) := \left\{a^c : c \in \mathbb{F}_{q^m}^*\right\}$ and $\mathcal{C}(\gamma)$ is called the trivial conjugacy class. There are $q-1$ distinct nontrivial (θ, δ)-conjugacy classes. In the zero-derivation case, each of the first $q-1$ powers of any primitive element of $\mathbb{F}_{q^m}$ belongs to another nontrivial class.

2.1 Skew-Polynomial Rings

Skew polynomials were first studied by Ore in 1933 [34,35] and are used e.g. for the construction of LRS codes [29]. The skew-polynomial ring $\mathbb{F}_{q^m}[x; \theta, \delta]$ contains all formal polynomials $\sum_i f_i x^{i-1}$ with finitely many nonzero coefficients $f_i \in \mathbb{F}_{q^m}$. The notion of the degree $\deg(f) := \max\{i-1 : f_i \neq 0\}$ of a skew polynomial $f(x) = \sum_i f_i x^{i-1}$ carries over from $\mathbb{F}_{q^m}[x]$. The set of skew polynomials forms a non-commutative ring with respect to conventional polynomial addition and a multiplication that is determined by the non-commutative rule $xa = \theta(a)x + \delta(a)$ for any $a \in \mathbb{F}_{q^m}$. By $\mathbb{F}_{q^m}[x;\theta,\delta]_{<k}$ we denote the subset of $\mathbb{F}_{q^m}[x;\theta,\delta]$ containing all skew polynomials of degree less than k. For simplicity, we refer to the skew-polynomial ring with zero derivation by $\mathbb{F}_{q^m}[x;\theta] := \mathbb{F}_{q^m}[x;\theta,0]$.

$\mathbb{F}_{q^m}[x;\theta,\delta]$ is Euclidean which ensures the existence of skew polynomials $q, r \in \mathbb{F}_{q^m}[x;\theta,\delta]$ with $f(x) = q(x)g(x) + r(x)$ and $\deg(r) < \deg(g)$ for each pair $f, g \in \mathbb{F}_{q^m}[x;\theta,\delta]$ with $\deg(f) \geq \deg(g)$. We denote the remainder r of this right-hand division by $f \bmod_{\mathrm{r}} g$.

The literature provides two meaningful ways to evaluate skew polynomials, namely, the remainder evaluation [25] and the generalized operator evaluation [29]. The former corresponds to the idea of enforcing a remainder theorem similar to the one in conventional polynomial rings and will not be of interest for this paper. The latter is e.g. used for the construction of LRS codes that we heavily rely on. For defining the generalized operator evaluation of skew polynomials we first introduce the operator $\mathcal{D}_a(b) := \theta(b)a + \delta(b)$ and its i-th power $\mathcal{D}_a^i(b) := \mathcal{D}_a(\mathcal{D}_a^{i-1}(b))$ for $i \in \mathbb{N}^*$ and any $a, b \in \mathbb{F}_{q^m}$. The operator simplifies to $\mathcal{D}_a(b) = \theta(b)a$ for all $a, b \in \mathbb{F}_{q^m}$ in the case of zero derivation. In this case, its i-th power $\mathcal{D}_a^i(b)$ for $i \in \mathbb{N}^*$ can be written as $\mathcal{D}_a^i(b) = \theta^i(b) \cdot \mathcal{N}_i(a)$, where $\mathcal{N}_i(a) := \prod_{k=0}^{i-1} \theta^k(a)$ is the i-th truncated norm of a.

The *generalized operator evaluation* of a skew polynomial $f(x) = \sum_{i=1}^{d} f_i x^{i-1} \in \mathbb{F}_{q^m}[x;\theta,\delta]$ at a point $b \in \mathbb{F}_{q^m}$ and with respect to an evaluation parameter $a \in \mathbb{F}_{q^m}$ is defined as

$$f(b)_a := \sum_{i=1}^{d} f_i \mathcal{D}_a^{i-1}(b).$$

We use the notation $f(\boldsymbol{b})_a := (f(b_1)_a, \ldots, f(b_n)_a)$ to denote the vector containing the evaluations of f at every entry of $\boldsymbol{b} \in \mathbb{F}_{q^m}^n$. Moreover, if $\boldsymbol{b} = (\boldsymbol{b}^{(1)} \mid \cdots \mid \boldsymbol{b}^{(\ell)}) \in \mathbb{F}_{q^m}^n$ is subdivided according to a length partition $\boldsymbol{n}$ and $\boldsymbol{a} = (a_1, \ldots, a_\ell) \in \mathbb{F}_{q^m}^\ell$, we use the shorthand $f(\boldsymbol{b})_{\boldsymbol{a}} := (f(\boldsymbol{b}^{(1)})_{a_1}, \ldots, f(\boldsymbol{b}^{(\ell)})_{a_\ell})$ to evaluate f at the elements of the i-th block $\boldsymbol{b}^{(i)}$ with respect to the evaluation parameter a_i for every $i = 1, \ldots, \ell$.

The evaluation of a product of two skew polynomials $f, g \in \mathbb{F}_{q^m}[x;\theta,\delta]$ satisfies the product rule $(f \cdot g)(b)_a = f(g(b)_a)_a$ for all $a, b \in \mathbb{F}_{q^m}$ [25].

For a vector $\boldsymbol{x} = (\boldsymbol{x}^{(1)} \mid \cdots \mid \boldsymbol{x}^{(\ell)}) \in \mathbb{F}_{q^m}^n$, a vector $\boldsymbol{a} \in \mathbb{F}_{q^m}^\ell$, and a parameter $d \in \mathbb{N}^*$ the *generalized Moore matrix* $\mathfrak{M}_d(\boldsymbol{x})_{\boldsymbol{a}}$ is defined as

$$\mathfrak{M}_d(\boldsymbol{x})_{\boldsymbol{a}} := \left(\mathfrak{m}_d(\boldsymbol{x}^{(1)})_{a_1} \mid \cdots \mid \mathfrak{m}_d(\boldsymbol{x}^{(\ell)})_{a_\ell}\right) \in \mathbb{F}_{q^m}^{d \times n},$$

$$\text{with } \mathfrak{m}_d(\boldsymbol{x}^{(i)})_{a_i} := \begin{pmatrix} x_1^{(i)} & \cdots & x_{n_i}^{(i)} \\ \mathcal{D}_{a_i}(x_1^{(i)}) & \cdots & \mathcal{D}_{a_i}(x_{n_i}^{(i)}) \\ \vdots & \ddots & \vdots \\ \mathcal{D}_{a_i}^{d-1}(x_1^{(i)}) & \cdots & \mathcal{D}_{a_i}^{d-1}(x_{n_i}^{(i)}) \end{pmatrix} \quad \text{for all } i = 1, \ldots, \ell.$$

If $\boldsymbol{a}$ contains representatives of pairwise distinct nontrivial conjugacy classes of $\mathbb{F}_{q^m}$ and $\mathrm{rk}_q\left(\boldsymbol{x}^{(i)}\right) = n_i$ for all $i = 1, \ldots, \ell$, it holds $\mathrm{rk}_{q^m}\left(\mathfrak{M}_d(\boldsymbol{x})_{\boldsymbol{a}}\right) = \min(d, n)$ [25,29].

Consider $\boldsymbol{b} = (\boldsymbol{b}^{(1)} \mid \cdots \mid \boldsymbol{b}^{(\ell)}) \in \mathbb{F}_{q^m}^n$ and $\boldsymbol{a} = (a_1, \ldots, a_\ell) \in \mathbb{F}_{q^m}^\ell$. The minimal skew polynomial that vanishes on the entries of $\boldsymbol{b}^{(i)}$ with respect to the evaluation parameter a_i for each $i = 1, \ldots, \ell$ is denoted by $\mathrm{mpol}_{(\boldsymbol{b})_{\boldsymbol{a}}}(x)$ and characterized by

$$\mathrm{mpol}_{(\boldsymbol{b})_{\boldsymbol{a}}}(\boldsymbol{b}^{(i)})_{a_i} = \boldsymbol{0} \quad \text{for all } i = 1, \ldots, \ell.$$

According to [15], it can be computed as a least common left multiple (lclm) via

$$\mathrm{mpol}_{(\boldsymbol{b})_{\boldsymbol{a}}}(x) = \mathrm{lclm}\left\{ x - \frac{\mathcal{D}_{a_i}(b_\iota^{(i)})}{b_\iota^{(i)}} \;:\; b_\iota^{(i)} \neq 0, \quad \begin{array}{c} \iota = 1, \ldots, n_i, \\ i = 1, \ldots, \ell \end{array} \right\}. \tag{1}$$

The degree satisfies $\deg(\mathrm{mpol}_{(\boldsymbol{b})_{\boldsymbol{a}}}) \leq n$ with equality if and only if the entries of $\boldsymbol{b}^{(i)}$ are $\mathbb{F}_q$-linearly independent for all $i = 1, \ldots, \ell$ and the evaluation parameters $a_1, \ldots, a_\ell$ belong to distinct nontrivial conjugacy classes of $\mathbb{F}_{q^m}$.

Now consider an additional vector $\boldsymbol{c} = (\boldsymbol{c}^{(1)} \mid \cdots \mid \boldsymbol{c}^{(\ell)}) \in \mathbb{F}_{q^m}^n$. Then there exists a unique skew interpolation polynomial $\mathrm{intpol}^{\boldsymbol{c}}_{(\boldsymbol{b})_{\boldsymbol{a}}}(x) \in \mathbb{F}_{q^m}[x; \theta, \delta]$ with $\deg(\mathrm{intpol}^{\boldsymbol{c}}_{(\boldsymbol{b})_{\boldsymbol{a}}}) < n$ and

$$\mathrm{intpol}^{\boldsymbol{c}}_{(\boldsymbol{b})_{\boldsymbol{a}}}(\boldsymbol{b}^{(i)})_{a_i} = \boldsymbol{c}^{(i)} \quad \text{for all } i = 1, \ldots, \ell \text{ [16]}.$$

For the complexity analysis of the Gao-like decoder, we will use $\mathcal{O}(\cdot)$ to state asymptotic costs in terms of the usual big-O notation. Moreover, the notation $\tilde{\mathcal{O}}(\cdot)$ indicates that logarithmic factors in the input parameter are neglected. The complexity of skew-polynomial operations in the zero-derivation setting was summarized in [11, Section II.D.]. Particularly, left and right division of skew polynomials with degree at most n as well as the computation of a minimal or an interpolation polynomial of degree at most n can be achieved in $\tilde{\mathcal{O}}(\mathcal{M}_{q,m}(n))$ operations in $\mathbb{F}_{q^m}$. Here, $\mathcal{M}_{q,m}(n)$ denotes the cost of multiplying two skew polynomials of degree n from $\mathbb{F}_{q^m}[x; \theta]$ and it holds $\mathcal{O}(\mathcal{M}_{q,m}(n)) \subseteq \mathcal{O}(n^{\min(\frac{\omega+1}{2}, 1.635)}) \subseteq \mathcal{O}(n^{1.635})$. The exponent $\omega \geq 2$ denotes the matrix-multiplication coefficient for which the currently best known upper bound is $\omega < 2.3728639$ [26].

2.2 The Sum-Rank Metric and the Corresponding Interleaved Channel Model

The *sum-rank weight* of a vector $\boldsymbol{x} = (\boldsymbol{x}^{(1)} \mid \cdots \mid \boldsymbol{x}^{(\ell)}) \in \mathbb{F}_{q^m}^n$ with respect to the length partition $\boldsymbol{n}$ is

$$\mathrm{wt}_{\Sigma R, \boldsymbol{n}}(\boldsymbol{x}) = \sum_{i=1}^{\ell} \mathrm{rk}_q\left(\boldsymbol{x}^{(i)}\right)$$

where $\mathrm{rk}_q\left(\boldsymbol{x}^{(i)}\right)$ is the maximum number of $\mathbb{F}_q$-linearly independent entries of the block $\boldsymbol{x}^{(i)}$ for each $i = 1, \ldots, \ell$. The *sum-rank metric* is induced by the sum-rank weight via $d_{\Sigma R,\boldsymbol{n}}(\boldsymbol{x}, \boldsymbol{y}) = \mathrm{wt}_{\Sigma R,\boldsymbol{n}}(\boldsymbol{x} - \boldsymbol{y})$ for all vectors $\boldsymbol{x}, \boldsymbol{y} \in \mathbb{F}_{q^m}^n$. Note that we omit the index $\boldsymbol{n}$ and simply write $\mathrm{wt}_{\Sigma R}$ and $d_{\Sigma R}$ when the length partition is clear from the context.

The sum-rank metric coincides with the Hamming metric for $\ell = n$, i.e., when every block has length one, and with the rank metric for $\ell = 1$, i.e., when the vector is considered as a single block.

Let now $\boldsymbol{x} = (\boldsymbol{x}_1 \mid \cdots \mid \boldsymbol{x}_s) \in \mathbb{F}_{q^m}^{sn}$ with $\boldsymbol{x}_j \in \mathbb{F}_{q^m}^n$ for all $j = 1, \ldots, s$ be a horizontally s-interleaved vector for an interleaving order $s \in \mathbb{N}^*$. Let us further assume for simplicity that all component vectors $\boldsymbol{x}_j = (\boldsymbol{x}_j^{(1)} \mid \cdots \mid \boldsymbol{x}_j^{(\ell)}) \in \mathbb{F}_{q^m}^n$ for $j = 1, \ldots, s$ are equipped with the same length partition $\boldsymbol{n}$. The natural way to define the sum-rank weight of $\boldsymbol{x} \in \mathbb{F}_{q^m}^{sn}$ is with respect to the *block-ordered* length partition $\tilde{\boldsymbol{n}} = (sn_1, \ldots, sn_\ell)$, i.e., as

$$\mathrm{wt}_{\Sigma R,\tilde{\boldsymbol{n}}}(\boldsymbol{x}) := \sum_{i=1}^{\ell} \mathrm{rk}_q(\boldsymbol{x}^{(i)}) \qquad \text{for } \boldsymbol{x}^{(i)} = (\boldsymbol{x}_1^{(i)} \mid \cdots \mid \boldsymbol{x}_s^{(i)}).$$

As for the conventional sum-rank metric, we often omit the length partition in the index and simply write $\mathrm{wt}_{\Sigma R}(\boldsymbol{x})$ when $\tilde{\boldsymbol{n}}$ is clear from the context. Figure 1 illustrates how the sum-rank weight of horizontally interleaved vectors is computed by grouping the same-indexed blocks of the component vectors. It shows how the block-ordered length partition arises naturally in this setting.

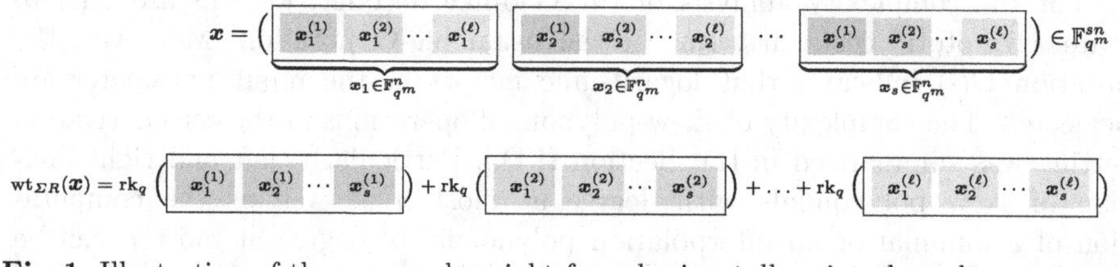

Fig. 1. Illustration of the sum-rank weight for a horizontally s-interleaved vector $\boldsymbol{x} = (\boldsymbol{x}_1 \mid \cdots \mid \boldsymbol{x}_s) \in \mathbb{F}_{q^m}^{sn}$.

We now consider the transmission of an interleaved vector $\boldsymbol{x} \in \mathbb{F}_{q^m}^{sn}$ over a sum-rank error channel with output

$$\boldsymbol{y} = \boldsymbol{x} + \boldsymbol{e} \tag{2}$$

where the error vector $\boldsymbol{e}$ is understood as a horizontally s-interleaved vector $\boldsymbol{e} = (\boldsymbol{e}_1 \mid \cdots \mid \boldsymbol{e}_s) \in \mathbb{F}_{q^m}^{sn}$ of sum-rank weight $\mathrm{wt}_{\Sigma R,\tilde{\boldsymbol{n}}}(\boldsymbol{e}) = t$. We further assume a uniform channel distribution, that is, that the error $\boldsymbol{e}$ is drawn uniformly at random from the set

$$\{\boldsymbol{x} = (\boldsymbol{x}_1 \mid \cdots \mid \boldsymbol{x}_s) \in \mathbb{F}_{q^m}^{sn} : \mathrm{wt}_{\Sigma R,\tilde{\boldsymbol{n}}}(\boldsymbol{x}) = t\}. \tag{3}$$

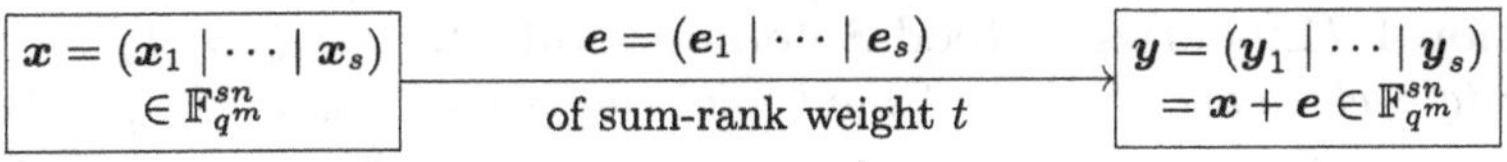

Fig. 2. The additive sum-rank channel for horizontally interleaved vectors.

The described channel is illustrated in Fig. 2.

Let $\boldsymbol{t} = (t_1, \ldots, t_\ell) \in \mathbb{N}^\ell$ with $t_i = \mathrm{rk}_q(\boldsymbol{e}^{(i)}) := \mathrm{rk}_q(\boldsymbol{e}_1^{(i)} \mid \cdots \mid \boldsymbol{e}_s^{(i)})$ for all $i = 1, \ldots, \ell$ denote the rank partition of $\boldsymbol{e}$. Then, we obtain for each $i = 1, \ldots, \ell$ a decomposition of the form $(\boldsymbol{e}_1^{(i)} \mid \cdots \mid \boldsymbol{e}_s^{(i)}) = \boldsymbol{a}^{(i)} \cdot \left(\boldsymbol{B}_1^{(i)} \mid \cdots \mid \boldsymbol{B}_s^{(i)}\right)$, where $\boldsymbol{a}^{(i)} \in \mathbb{F}_{q^m}^{t_i}$ with $\mathrm{rk}_q(\boldsymbol{a}^{(i)}) = t_i$ and $\boldsymbol{B}_j^{(i)} \in \mathbb{F}_q^{t_i \times n_i}$ with $\mathrm{rk}_q\left(\boldsymbol{B}_1^{(i)} \mid \cdots \mid \boldsymbol{B}_s^{(i)}\right) = t_i$ for all $j = 1, \ldots, s$. After reordering the components, the error vector $\boldsymbol{e}$ can thus be decomposed as

$$\boldsymbol{e} = \boldsymbol{a} \cdot \boldsymbol{B} \tag{4}$$

with $\boldsymbol{a} = (\boldsymbol{a}^{(1)} \mid \cdots \mid \boldsymbol{a}^{(\ell)}) \in \mathbb{F}_{q^m}^{t}$ and

$$\boldsymbol{B} = \left(\begin{array}{ccc|c|ccc} \boldsymbol{B}_1^{(1)} & & & & \boldsymbol{B}_s^{(1)} & & \\ & \ddots & & \cdots & & \ddots & \\ & & \boldsymbol{B}_1^{(\ell)} & & & & \boldsymbol{B}_s^{(\ell)} \end{array}\right) \in \mathbb{F}_q^{t \times sn}, \tag{5}$$

where for any $i = 1, \ldots, \ell$ and any $j = 1, \ldots, s$

$$\boldsymbol{a}^{(i)} \in \mathbb{F}_{q^m}^{t_i} \text{ with } \mathrm{rk}_q(\boldsymbol{a}^{(i)}) = t_i$$
$$\text{and } \boldsymbol{B}_j^{(i)} \in \mathbb{F}_q^{t_i \times n_i} \text{ with } \mathrm{rk}_q\left(\boldsymbol{B}_1^{(i)} \mid \cdots \mid \boldsymbol{B}_s^{(i)}\right) = t_i.$$

Note that the decomposition in (4) is not unique. Moreover, the uniform distribution of $\boldsymbol{e}$ among all vectors of sum-rank weight t implies that, for fixed rank partition $\boldsymbol{t}$, both $\boldsymbol{a}$ and $\boldsymbol{B}$ are also chosen uniformly at random from the sets

$$\{\boldsymbol{x} \in \mathbb{F}_{q^m}^{t} : \mathrm{wt}_{\Sigma R, \boldsymbol{t}}(\boldsymbol{x}) = t\}$$
$$\text{and} \quad \{\boldsymbol{X} \in \mathbb{F}_{q^m}^{t \times sn} \text{ of the form (5)} : \mathrm{wt}_{\Sigma R, \tilde{\boldsymbol{n}}}(\boldsymbol{X}) = t\}, \tag{6}$$

respectively.

The elements in $\boldsymbol{a}^{(i)}$ form a basis of the column space of $\boldsymbol{e}^{(i)}$ and are called *error values*. Similarly, the rows of $\boldsymbol{B}_j^{(i)}$ form a basis of the row space of $\boldsymbol{e}_j^{(i)}$ and are referred to as *error locations*. For horizontal interleaving, the error values in $\boldsymbol{a}$ are common for all component errors.

2.3 Horizontally Interleaved Linearized Reed–Solomon (HILRS) Codes

We first introduce LRS codes [29, Definition 31], which are one of the most prominent families of sum-rank-metric codes.

Definition 1 (Linearized Reed-Solomon Codes). *Let* $\boldsymbol{\xi} = (\xi_1, \dots, \xi_\ell) \in \mathbb{F}_{q^m}^\ell$ *contain elements of distinct nontrivial conjugacy classes of* $\mathbb{F}_{q^m}$. *Further denote by* $\boldsymbol{n} = (n_1, \dots, n_\ell) \in \mathbb{N}^\ell$ *a length partition of* n, *i.e.,* $n = \sum_{i=1}^{\ell} n_i$. *Let the vectors* $\boldsymbol{\beta}^{(i)} = (\beta_1^{(i)}, \dots, \beta_{n_i}^{(i)}) \in \mathbb{F}_{q^m}^{n_i}$ *contain* $\mathbb{F}_q$*-linearly independent* $\mathbb{F}_{q^m}$*-elements for all* $i = 1, \dots, \ell$ *and write* $\boldsymbol{\beta} = \left(\boldsymbol{\beta}^{(1)} \mid \cdots \mid \boldsymbol{\beta}^{(\ell)}\right) \in \mathbb{F}_{q^m}^n$. *A linearized Reed–Solomon (LRS) code of length* n *and dimension* k *is defined as*

$$\mathrm{LRS}[\boldsymbol{\beta}, \boldsymbol{\xi}; \boldsymbol{n}, k] = \left\{ \left(f(\boldsymbol{\beta}^{(1)})_{\xi_1} \mid \cdots \mid f(\boldsymbol{\beta}^{(\ell)})_{\xi_\ell} \right) : f \in \mathbb{F}_{q^m}[x; \theta, \delta]_{<k} \right\} \subseteq \mathbb{F}_{q^m}^n.$$

Every codeword $\boldsymbol{c} \in \mathrm{LRS}[\boldsymbol{\beta}, \boldsymbol{\xi}; \boldsymbol{n}, k]$ corresponds to a skew polynomial $f \in \mathbb{F}_{q^m}[x; \theta, \delta]_{<k}$. We sometimes write $\boldsymbol{c} = \boldsymbol{c}(f)$ to emphasize this and call f the *message polynomial* of $\boldsymbol{c}$.

The minimum distance d of an LRS code satisfies the Singleton-like bound $d \leq n - k + 1$ with equality. Thus, LRS codes are maximum sum-rank distance (MSRD) codes.

Similar to RS and Gabidulin codes, LRS codes have a generator matrix $\boldsymbol{G}$ of a particularly useful form. Namely, the matrix $\boldsymbol{G} = (\boldsymbol{G}^{(1)} \mid \cdots \mid \boldsymbol{G}^{(\ell)}) = \mathfrak{M}_k(\boldsymbol{\beta})_{\boldsymbol{\xi}} \in \mathbb{F}_{q^m}^{k \times n}$ with

$$\boldsymbol{G}^{(i)} = \mathfrak{m}_k(\boldsymbol{\beta}^{(i)})_{\xi_i} = \begin{pmatrix} \beta_1^{(i)} & \cdots & \beta_{n_i}^{(i)} \\ \mathcal{D}_{\xi_i}(\beta_1^{(i)}) & \cdots & \mathcal{D}_{\xi_i}(\beta_{n_i}^{(i)}) \\ \vdots & \ddots & \vdots \\ \mathcal{D}_{\xi_i}^{k-1}(\beta_1^{(i)}) & \cdots & \mathcal{D}_{\xi_i}^{k-1}(\beta_{n_i}^{(i)}) \end{pmatrix} \in \mathbb{F}_{q^m}^{k \times n_i}$$

for all $i = 1, \dots, \ell$ generates the code $\mathrm{LRS}[\boldsymbol{\beta}, \boldsymbol{\xi}; \boldsymbol{n}, k]$.

We obtain an HILRS code with interleaving order $s \in \mathbb{N}^*$ by combining s LRS component codes. Namely, each codeword of the HILRS code is the horizontal concatenation of s codewords of the chosen component codes.

Definition 2 (Horizontally Interleaved LRS Codes). *Fix an interleaving order* $s \in \mathbb{N}^*$ *and pick for each* $j = 1, \dots, s$ *an LRS code* $\mathrm{LRS}[\boldsymbol{\beta}_j, \boldsymbol{\xi}; \boldsymbol{n}, k]$ *according to Definition 1. We define the horizontally interleaved linearized Reed–Solomon (HILRS) code with interleaving order* s, *code locators* $\boldsymbol{\beta} := (\boldsymbol{\beta}_1 \mid \cdots \mid \boldsymbol{\beta}_s)$, *evaluation parameters* $\boldsymbol{\xi}$, *and length partition* $s\boldsymbol{n} := (sn_1, \dots, sn_\ell)$ *as*

$$\mathrm{HILRS}[\boldsymbol{\beta}, \boldsymbol{\xi}, s; s\boldsymbol{n}, sk] = \left\{ (\boldsymbol{c}_1 \mid \cdots \mid \boldsymbol{c}_s) : \begin{array}{c} \boldsymbol{c}_j \in \mathrm{LRS}[\boldsymbol{\beta}_j, \boldsymbol{\xi}; \boldsymbol{n}, k] \\ \textit{for all } j = 1, \dots, s \end{array} \right\} \subseteq \mathbb{F}_{q^m}^{sn}.$$

The code $\mathrm{HILRS}[\boldsymbol{\beta}, \boldsymbol{\xi}, s; s\boldsymbol{n}, sk]$ has length sn and dimension sk over $\mathbb{F}_{q^m}$. Its minimum distance d equals the minimum distance of its component codes, i.e., $d = n - k + 1$. HILRS codes are hence *not* MSRD. Similar to LRS codes, we write $\boldsymbol{c}(\boldsymbol{f}) = (\boldsymbol{c}_1(f_1) \mid \cdots \mid \boldsymbol{c}_s(f_s)) \in \mathrm{HILRS}[\boldsymbol{\beta}, \boldsymbol{\xi}, s; s\boldsymbol{n}, sk]$ with $\boldsymbol{f} = (f_1, \dots, f_s)$ and $f_j \in \mathbb{F}_{q^m}[x; \theta, \delta]_{<k}$ for each $j = 1, \dots, s$ to emphasize the relation to the message polynomials of the component codewords $\boldsymbol{c}_1, \dots, \boldsymbol{c}_s$. We call $\boldsymbol{f}$ the *message-polynomial vector* corresponding to $\boldsymbol{c}$.

Remark 1. It is straightforward to generalize Definition 2 and all concepts of this paper to component codes with different length partitions, lengths, and dimensions. However, we assume that the component codes only have different code locators $\boldsymbol{\beta}_j$ for $j = 1, \ldots, s$ for simplicity of notation. □

3 A Gao-Like Decoder for HILRS Codes

We now derive a Gao-like decoder in the spirit of [20,36,45] for HILRS codes and the interleaved sum-rank-channel model described in (2). Let $\boldsymbol{y} = \boldsymbol{c} + \boldsymbol{e} \in \mathbb{F}_{q^m}^{sn}$ denote the received vector after the codeword $\boldsymbol{c} = \boldsymbol{c}(\boldsymbol{f}) \in \mathrm{HILRS}[\boldsymbol{\beta}, \boldsymbol{\xi}, s; \boldsymbol{sn}, sk]$ was corrupted by the error $\boldsymbol{e} \in \mathbb{F}_{q^m}^{sn}$ of sum-rank weight $\mathrm{wt}_{\Sigma R}(\boldsymbol{e}) = t$ during transmission. Recall that we assume a uniform error distribution, that is, that $\boldsymbol{e}$ is chosen uniformly at random from the set of all vectors of sum-rank weight t as given in (3).

The main ingredient of the decoder is the Gao-like key equation that exploits the relation between certain polynomials to recover the error values as zeros of the error-span polynomial. Then, the message-polynomial vector $\boldsymbol{f}$ that corresponds to $\boldsymbol{c}$ can be retrieved.

The *error span polynomial (ESP)* $\sigma \in \mathbb{F}_{q^m}[x; \theta, \delta]$ makes use of the error decomposition shown in (4). It is the skew polynomial that vanishes at all error values, i.e.,

$$\sigma(\boldsymbol{a}^{(i)})_{\xi_i} = \boldsymbol{0} \qquad \text{for all } i = 1, \ldots, \ell.$$

For horizontal interleaving, the component errors $\boldsymbol{e}_j$ share the same error values $\boldsymbol{a}$ for all $j = 1, \ldots, s$ according to (4). This implies that the ESP is common for all component errors.

Next let $G_j \in \mathbb{F}_{q^m}[x; \theta, \delta]$ for each $j = 1, \ldots, s$ be the minimal skew polynomial for the code locators $\boldsymbol{\beta}_j$ with respect to generalized operator evaluation. Namely,

$$G_j(x) := \mathrm{mpol}_{(\boldsymbol{\beta}_j)_{\boldsymbol{\xi}}}(x) \qquad \text{for all } j = 1, \ldots, s.$$

Remark that these polynomials only depend on code parameters and can thus be precomputed. Further, define $R_j \in \mathbb{F}_{q^m}[x; \theta, \delta]$ for each $j = 1, \ldots, s$ as the interpolation polynomial whose evaluation at the code locators $\boldsymbol{\beta}_j$ yields the channel observation $\boldsymbol{y}_j$. That means that $R_j(x) := \mathrm{intpol}^{\boldsymbol{y}_j}_{(\boldsymbol{\beta}_j)_{\boldsymbol{\xi}}}(x)$ satisfies

$$R_j(\boldsymbol{\beta}_j)_{\boldsymbol{\xi}} = \boldsymbol{y}_j \qquad \text{for all } j = 1, \ldots, s.$$

Note that the polynomials R_j can be computed directly from the channel observation $\boldsymbol{y} = (\boldsymbol{y}_1 \mid \cdots \mid \boldsymbol{y}_s)$.

Theorem 1 (Gao-like Key Equation for HILRS Codes). *Let* $\mathbf{c} = \boldsymbol{c}(\boldsymbol{f}) \in \mathrm{HILRS}[\boldsymbol{\beta}, \boldsymbol{\xi}, s; \boldsymbol{sn}, sk]$ *be a codeword corresponding to the message-polynomial vector* $\boldsymbol{f} = (f_1, \ldots, f_s)$ *with* $f_j \in \mathbb{F}_{q^m}[x; \theta, \delta]_{<k}$ *for all* $j = 1, \ldots, s$. *Let further* $\boldsymbol{y} = \boldsymbol{c} + \boldsymbol{e} \in \mathbb{F}_{q^m}^{sn}$ *denote a channel observation according to* (2). *For the ESP* $\sigma \in \mathbb{F}_{q^m}[x; \theta, \delta]$ *and the polynomials*

$$G_j(x) = \mathrm{mpol}_{(\boldsymbol{\beta}_j)_{\boldsymbol{\xi}}}(x) \quad \text{and} \quad R_j(x) = \mathrm{intpol}^{\boldsymbol{y}_j}_{(\boldsymbol{\beta}_j)_{\boldsymbol{\xi}}}(x) \qquad \text{for each } j = 1, \ldots, s,$$

it holds

$$\sigma \cdot R_j \equiv \sigma \cdot f_j \bmod_r G_j \quad \textit{for all } j = 1, \ldots, s. \tag{7}$$

Proof. Consider a fixed $j = 1, \ldots, s$ and let us show the equivalent formulation

$$\sigma \cdot (R_j - f_j) \equiv 0 \bmod_r G_j$$

of the key equation. By definition, we know that the evaluation of $R_j - f_j$ at $\boldsymbol{\beta}_j$ is $(R_j - f_j)(\boldsymbol{\beta}_j)_{\xi} = \boldsymbol{y}_j - \boldsymbol{c}_j = \boldsymbol{e}_j$. Thus,

$$(\sigma \cdot (R_j - f_j))(\boldsymbol{\beta}_j)_{\xi} \overset{(\triangle)}{=} \sigma((R_j - f_j)(\boldsymbol{\beta}_j)_{\xi})_{\xi} = \sigma(\boldsymbol{e}_j)_{\xi} = \boldsymbol{0}$$

applies, where $(\triangle)$ follows from the product rule for generalized operator evaluation and the other equalities hold by definition. Together with the fact that G_j is the minimal polynomial of the code locators, we conclude that G_j divides $\sigma \cdot (R_j - f_j)$ on the right. Since this argument is true for every $j = 1, \ldots, s$, the statement follows. □

As can be seen from the proof of Theorem 1, the Gao-like key Eq. (7) is in fact equivalent to

$$(\sigma \cdot (R_j - f_j))(\boldsymbol{\beta}_j)_{\xi} = \boldsymbol{0} \qquad \text{for all } j = 1, \ldots, s.$$

By rewriting it in terms of a system of $\mathbb{F}_{q^m}$-linear equations, we obtain

$$\underbrace{\begin{pmatrix} (\mathfrak{M}_{t+k}(\boldsymbol{\beta}_1)_{\xi})^{\top} & & -(\mathfrak{M}_{t+1}(\boldsymbol{y}_1)_{\xi})^{\top} \\ & \ddots & \vdots \\ & (\mathfrak{M}_{t+k}(\boldsymbol{\beta}_s)_{\xi})^{\top} & -(\mathfrak{M}_{t+1}(\boldsymbol{y}_s)_{\xi})^{\top} \end{pmatrix}}_{=:\boldsymbol{M}^{\top}} \cdot \begin{pmatrix} \boldsymbol{\sigma f}_1 \\ \vdots \\ \boldsymbol{\sigma f}_s \\ \boldsymbol{\sigma} \end{pmatrix} = \boldsymbol{0}. \tag{8}$$

Here, the vectors $\boldsymbol{\sigma}$ and $\boldsymbol{\sigma f}_j$ for $j = 1, \ldots, s$ contain the coefficients of the respective polynomials, i.e.,

$$(\boldsymbol{\sigma f}_j)^{\top} := ((\sigma \cdot f_j)_1, \ldots, (\sigma \cdot f_j)_{t+k}) \in \mathbb{F}_{q^m}^{t+k} \quad \text{for all } j = 1, \ldots, s$$
$$\text{and} \qquad \boldsymbol{\sigma}^{\top} := (\sigma_1, \ldots, \sigma_{t+1}) \in \mathbb{F}_{q^m}^{t+1}.$$

Equation (8) displays a homogeneous system of sn equations in $s(t+k)+t+1 = (s+1)t + sk + 1$ unknowns. It can be solved by Gaussian elimination with a complexity of $\mathcal{O}(\max(sn, (s+1)t + sk + 1)^{\omega})$ operations in $\mathbb{F}_{q^m}$ [44, Proposition 2.15.].

As soon as the Gao-like key equation is solved, we have access to a candidate $\tilde{\sigma}$ for the ESP $\sigma \in \mathbb{F}_{q^m}[x; \theta, \delta]$ as well as to candidates p_j for the products $\sigma \cdot f_j \in \mathbb{F}_{q^m}[x; \theta, \delta]_{<t+k}$ for all $j = 1, \ldots, s$. Thus, for any $j = 1, \ldots, s$, left division of p_j by $\tilde{\sigma}$ recovers a candidate $\tilde{f}_j$ for the j-th message polynomial f_j. If the remainder r_j of the left division of p_j by $\tilde{\sigma}$ is nonzero for any $j = 1, \ldots, s$ or if any of the $\tilde{f}_1, \ldots, \tilde{f}_s$ has degree at least k, we declare a decoding failure.

Algorithm 1: Gao-like Decoder for HILRS Codes

Input : received vector $\boldsymbol{y} \in \mathbb{F}_{q^m}^{sn}$ with $\boldsymbol{y} = \boldsymbol{c}(\boldsymbol{f}) + \boldsymbol{e}$ according to (2) and with $\boldsymbol{c}(\boldsymbol{f}) \in \mathrm{HILRS}[\boldsymbol{\beta}, \boldsymbol{\xi}, s; s\boldsymbol{n}, sk]$
precomputed $G_1, \ldots, G_s$ with $G_j := \mathrm{mpol}_{(\boldsymbol{\beta}_j)_{\boldsymbol{\xi}}}(x)$ for all $j = 1, \ldots, s$

Output : $\boldsymbol{f} = (f_1, \ldots, f_s)$ or "decoding failure"

$R_j := \mathrm{intpol}_{(\boldsymbol{\beta}_j)_{\boldsymbol{\xi}}}^{\boldsymbol{y}_j}(x) \in \mathbb{F}_{q^m}[x; \theta, \delta]$ for all $j = 1, \ldots, s$

/* use $\sigma \cdot R_j \equiv \sigma \cdot f_j \bmod_{\mathrm{r}} G_j$ to find $p_j \triangleq \sigma \cdot f_j$ and $\tilde{\sigma} \triangleq \sigma$ */

$(p_1, \ldots, p_s, \tilde{\sigma}) := \mathsf{solveKeyEquation}(R_1, \ldots, R_s, G_1, \ldots, G_s, n, k, s)$

forall $j = 1, \ldots, s$ **do**

  $(\tilde{f}_j, r_j) := \mathsf{leftDivide}(p_j, \tilde{\sigma})$

  if $r_j \neq 0$ **or** $\deg(\tilde{f}_j) \geq k$ **then**

    return "decoding failure"

return $\boldsymbol{f} := (\tilde{f}_1, \ldots, \tilde{f}_s)$

Otherwise, the decoding was correct and $\tilde{f}_j = f_j$ applies for all $j = 1, \ldots, s$. Algorithm 1 summarizes all steps of the Gao-like decoder.

Let us now further investigate the structure of $\boldsymbol{M}^\top$, which gives rise to the decoding-failure probability $\Pr_{\mathrm{fail}}$. Remark that the system (8) has a nontrivial solution by definition, which implies $\mathrm{rk}_{q^m}(\boldsymbol{M}) \leq (s+1)t + sk$. Moreover, a decoding failure can only occur if the solution space of (8) has dimension greater than one. In other words, $\mathrm{rk}_{q^m}(\boldsymbol{M}^\top) = \mathrm{rk}_{q^m}(\boldsymbol{M}) < (s+1)t + sk$ must apply and we obtain the inequality

$$\Pr_{\mathrm{fail}} \leq \Pr\left(\mathrm{rk}_{q^m}(\boldsymbol{M}) < (s+1)t + sk\right).$$

The following lemma gives a characterization of when the solution space of (8) is one-dimensional. Recall that this case implies correct decoding.

Lemma 1. *Consider a vector* $\boldsymbol{y} = \mathbf{c} + \boldsymbol{e} \in \mathbb{F}_{q^m}^{sn}$ *that was received after transmitting* $\mathbf{c} \in \mathrm{HILRS}[\boldsymbol{\beta}, \boldsymbol{\xi}, s; s\boldsymbol{n}, sk]$ *over the channel* (2). *Assume that the error has weight* $\mathrm{wt}_{\Sigma R}(\boldsymbol{e}) = t \leq n - k$ *and can be decomposed into* $\mathbf{e} = \boldsymbol{a} \cdot \boldsymbol{B}$ *according to* (4). *Further, define* $\boldsymbol{M}$ *as in* (8) *and let* $\boldsymbol{H} = \mathrm{diag}(\boldsymbol{H}_1, \ldots, \boldsymbol{H}_s) \in \mathbb{F}_{q^m}^{s(n-k-t) \times sn}$ *be a parity-check matrix of the code* $\mathrm{HILRS}[\boldsymbol{\beta}, \boldsymbol{\xi}, s; s\boldsymbol{n}, s(k+t)]$. *Then,*

$$\mathrm{rk}_{q^m}(\boldsymbol{M}) = (s+1)t + sk \qquad \textit{if and only if} \qquad \mathrm{rk}_{q^m}(\boldsymbol{B}\boldsymbol{H}^\top) = t.$$

Proof. First note that the upper part of $\boldsymbol{M}$ is a generator matrix of the code $\mathrm{HILRS}[\boldsymbol{\beta}, \boldsymbol{\xi}, s; s\boldsymbol{n}, s(k+t)]$. In other words, the j-th block on its diagonal generates $\mathrm{LRS}[\boldsymbol{\beta}_j, \boldsymbol{\xi}; \boldsymbol{n}, k+t]$ for all $j = 1, \ldots, s$. For any $j = 1, \ldots, s$, the additivity of the generalized operator evaluation yields $\mathfrak{M}_{t+1}(\boldsymbol{y}_j)_{\boldsymbol{\xi}} = \mathfrak{M}_{t+1}(\boldsymbol{c}_j)_{\boldsymbol{\xi}} + \mathfrak{M}_{t+1}(\boldsymbol{e}_j)_{\boldsymbol{\xi}}$. Further, $\boldsymbol{c}_j \in \mathrm{LRS}[\boldsymbol{\beta}_j, \boldsymbol{\xi}; \boldsymbol{n}, k] = \langle \mathfrak{M}_k(\boldsymbol{\beta}_j)_{\boldsymbol{\xi}} \rangle_{q^m}$ implies $\mathcal{D}_{\boldsymbol{\xi}}^{\iota}(\boldsymbol{c}_j) \in \langle \mathfrak{M}_{k+\iota}(\boldsymbol{\beta}_j)_{\boldsymbol{\xi}} \rangle_{q^m}$ for all $\iota = 1, \ldots, t$. We can hence consider the matrix

$$\widetilde{\boldsymbol{M}} = \left(\begin{array}{ccc} \mathfrak{M}_{t+k}(\boldsymbol{\beta}_1)_{\boldsymbol{\xi}} & & \\ & \ddots & \\ & & \mathfrak{M}_{t+k}(\boldsymbol{\beta}_s)_{\boldsymbol{\xi}} \\ \hline \mathfrak{M}_{t+1}(\boldsymbol{e}_1)_{\boldsymbol{\xi}} & \dots & \mathfrak{M}_{t+1}(\boldsymbol{e}_s)_{\boldsymbol{\xi}} \end{array}\right) =: \left(\frac{\boldsymbol{U}}{\boldsymbol{L}}\right)$$

which has the same $\mathbb{F}_{q^m}$-linear row space, and thus the same $\mathbb{F}_{q^m}$-rank, as $\boldsymbol{M}$. In the following, we denote the upper $s(t+k)$ rows of $\boldsymbol{M}$ by $\boldsymbol{U}$ and the lower part by $\boldsymbol{L}$ for convenience. The error decomposition and the $\mathbb{F}_q$-linearity of the generalized operator evaluation let us write $\boldsymbol{L} = \mathfrak{M}_{t+1}(\boldsymbol{a})_{\boldsymbol{\xi}} \cdot \boldsymbol{B}$. Therefore,

$$\widetilde{\boldsymbol{M}} = \left(\begin{array}{c|c} \boldsymbol{I}_{s(t+k)} & \boldsymbol{0} \\ \hline \boldsymbol{0} & \mathfrak{M}_{t+1}(\boldsymbol{a})_{\boldsymbol{\xi}} \end{array}\right) \cdot \left(\frac{\boldsymbol{U}}{\boldsymbol{B}}\right)$$

applies, where $\boldsymbol{I}_{s(t+k)}$ denotes the identity matrix of size $s(t+k) \times s(t+k)$. Since the left matrix has full column rank over $\mathbb{F}_{q^m}$, [31, Theorem 2] yields

$$\mathrm{rk}_{q^m}(\widetilde{\boldsymbol{M}}) = \mathrm{rk}_{q^m}\left(\frac{\boldsymbol{U}}{\boldsymbol{B}}\right).$$

Define $\boldsymbol{H} := \mathrm{diag}(\boldsymbol{H}_1, \dots, \boldsymbol{H}_s) \in \mathbb{F}_{q^m}^{s(n-k-t)\times sn}$ with $\boldsymbol{H}_j$ being a parity-check matrix of the code $\mathrm{LRS}[\boldsymbol{\beta}_j, \boldsymbol{\xi}; \boldsymbol{n}, k+t]$ for all $j = 1, \dots, s$. Then, $\boldsymbol{H}$ is a parity-check matrix of $\mathrm{HILRS}[\boldsymbol{\beta}, \boldsymbol{\xi}, s; s\boldsymbol{n}, s(k+t)]$ and satisfies $\boldsymbol{U}\boldsymbol{H}^\top = \boldsymbol{0}$. Since

$$\begin{aligned} \mathrm{rk}_{q^m}(\boldsymbol{M}) &= \mathrm{rk}_{q^m}(\boldsymbol{U}) + \mathrm{rk}_{q^m}(\boldsymbol{B}) - \dim_{q^m}(\langle \boldsymbol{U}\rangle_{q^m} \cap \langle \boldsymbol{B}\rangle_{q^m}) \\ &\leq (s+1)t + sk - \dim_{q^m}(\langle \boldsymbol{U}\rangle_{q^m} \cap \langle \boldsymbol{B}\rangle_{q^m}) \end{aligned}$$

holds, the equality $\mathrm{rk}_{q^m}(\boldsymbol{M}) = (s+1)t + sk$ is equivalent to $\langle \boldsymbol{U}\rangle_{q^m} \cap \langle \boldsymbol{B}\rangle_{q^m} = \{\boldsymbol{0}\}$ and thus to $\langle \boldsymbol{H}\rangle_{q^m}^{\perp} \cap \langle \boldsymbol{B}\rangle_{q^m} = \{\boldsymbol{0}\}$. This is equivalent to $\mathrm{rk}_{q^m}(\boldsymbol{B}\boldsymbol{H}^\top) = t$, which proves the lemma. □

This equivalent reformulation gives a condition on the error weight t and thus determines the decoding radius. In fact, the matrix $\boldsymbol{B}\boldsymbol{H}^\top$ has t rows and $s(n-k-t)$ columns and can achieve $\mathrm{rk}_{q^m}(\boldsymbol{B}\boldsymbol{H}^\top) = t$ only if $t \leq s(n-k-t)$ applies. Since we obtain a decoding failure in all other cases, we obtain the necessary condition

$$t \leq t_{\mathsf{max}} := \frac{s}{s+1}(n-k)$$

for successful decoding.

We now focus on the zero-derivation case and derive an upper bound on the probability that $\mathrm{rk}_{q^m}(\boldsymbol{B}\boldsymbol{H}^\top) < t$ which will also bound the decoding-failure probability according to Lemma 1. Recall that we can choose $\boldsymbol{H}$ such that $\boldsymbol{H}_1, \dots, \boldsymbol{H}_s$ are generalized Moore matrices, as the dual of an LRS code is again an LRS code in the zero-derivation setting [30, Theorem 4]. For such a choice of $\boldsymbol{H}$, the product $\boldsymbol{B}\boldsymbol{H}^\top = (\boldsymbol{B}_1\boldsymbol{H}_1^\top \mid \cdots \mid \boldsymbol{B}_s\boldsymbol{H}_s^\top)$ is the transpose of vertically stacked generalized Moore matrices because $\boldsymbol{B} = (\boldsymbol{B}_1 \mid \cdots \mid \boldsymbol{B}_s)$ contains only $\mathbb{F}_q$-elements and $\mathcal{D}_\xi(\cdot)$ is $\mathbb{F}_q$-linear for a fixed $\xi \in \mathbb{F}_{q^m}$. Namely,

$$\boldsymbol{H}\boldsymbol{B}^\top = \begin{pmatrix} \mathfrak{M}_{t+k}(\boldsymbol{h}_1\boldsymbol{B}_1^\top)_\xi \\ \dots \\ \mathfrak{M}_{t+k}(\boldsymbol{h}_s\boldsymbol{B}_s^\top)_\xi \end{pmatrix},$$

where $\boldsymbol{h}_j$ denotes the first row of $\boldsymbol{H}_j$ for each $j = 1, \dots, s$.

Further recall that, for a fixed rank partition $\boldsymbol{t}$, the matrix $\boldsymbol{B}$ is uniformly distributed among the set of all matrices of a particular form having fixed sum-rank weight as described in (6). As $\mathrm{wt}_{\Sigma R}(\boldsymbol{h}_j) = n$ applies for every $j = 1, \dots, s$, the $(s \times t)$-matrix containing the vectors $\boldsymbol{h}_j\boldsymbol{B}_j^\top$ as rows is chosen uniformly at random from all matrices in $\mathbb{F}_{q^m}^{s \times t}$ with sum-rank weight t. This allows us to apply parts of the proof of [13, Lemma 7].

In the zero-derivation setting, we thus obtain the upper bound

$$\Pr_{\mathsf{fail}} \leq \Pr\left(\mathrm{rk}_{q^m}(\boldsymbol{B}\boldsymbol{H}^\top) < t\right) \leq \kappa_q^{\ell+1} q^{-m((s+1)(t_{\max}-t)+1)} \tag{9}$$

on the decoding-failure probability $\Pr_{\mathsf{fail}}$, where $t_{\max} := \frac{s}{s+1}(n-k)$ and $\kappa_q < 3.5$ is defined as $\kappa_q := \prod_i \frac{1}{1-q^{-i}}$ for any prime power q.

We implemented the proposed decoder in SageMath [43] and ran a Monte Carlo simulation to heuristically verify the tightness of the upper bound on the decoding-failure probability given in (9). Note that the actual failure probability is hard to simulate for reasonable parameter sizes, as even the upper bound decreases exponentially. To obtain observable results, we chose $\mathbb{F}_{q^m} = \mathbb{F}_{3^8}$, $\mathbb{F}_q = \mathbb{F}_3$, and an HILRS code of length $n = 16$ and dimension $k = 4$ with respect to the Frobenius automorphism. We considered $\ell = 2$ blocks of the same length, namely $\boldsymbol{n} = (8, 8)$, interleaving order $s = 3$, and randomly chosen errors of sum-rank weight $t = t_{\max} = 9$. The failure probability that we observed for 100 Monte Carlo errors is $1.569 \cdot 10^{-4}$ while the bound yields $6.535 \cdot 10^{-3}$.

We finish this section with a summary of the results we have obtained so far and give a complexity analysis of the Gao-like decoder for HILRS codes.

Theorem 2 (Gao-like Decoding of HILRS Codes). *Consider the transmission of a codeword* $\mathbf{c} \in \mathrm{HILRS}[\boldsymbol{\beta}, \boldsymbol{\xi}, s; s\boldsymbol{n}, sk]$ *over the channel* (2). *Let* $\boldsymbol{y} = \boldsymbol{c} + \boldsymbol{e} \in \mathbb{F}_{q^m}^{sn}$ *denote the received word and assume that the error* $\mathbf{e}$ *has bounded sum-rank weight*

$$\mathrm{wt}_{\Sigma R}(\boldsymbol{e}) = t \leq \frac{s}{s+1}(n-k). \tag{10}$$

Then, the Gao-like decoder from Algorithm 1 can recover $\mathbf{c}$ *with a failure probability* $\Pr_{\mathsf{fail}}$ *that is bounded by*

$$\Pr_{\mathsf{fail}} \leq \kappa_q^{\ell+1} q^{-m((s+1)(t_{\max}-t)+1)} < 3.5^{\ell+1} q^{-m((s+1)(t_{\max}-t)+1)}$$

in the zero-derivation setting. If the key Eq. (7) *is solved via Gaussian elimination in the formulation of* (8), *the overall complexity of the decoder is in the order of* $\tilde{\mathcal{O}}((sn)^\omega) \subseteq \tilde{\mathcal{O}}((sn)^{2.373})$ *operations in* $\mathbb{F}_{q^m}$.

Proof. The decoding radius and the bound on the failure probability were derived above. Let us thus focus on the complexity analysis.

- The computation of a minimal or an interpolation polynomial of degree at most n can be done with complexity $\tilde{\mathcal{O}}(\mathcal{M}_{q,m}(n))$ according to [11, Section II.D.], e.g. by using the recursive formula (1). Thus, the computation of $G_1, \ldots, G_s$ and $R_1, \ldots, R_s$ takes $\tilde{\mathcal{O}}(s\mathcal{M}_{q,m}(n))$ operations in $\mathbb{F}_{q^m}$.
- Finding the solution of the key equation via Gaussian elimination has complexity $\mathcal{O}(\max(sn, (s+1)t + sk + 1)^\omega)$ as stated above. Since Eq. (10) ensures $sn \geq (s+1)t + sk + 1$, we obtain $\mathcal{O}((sn)^\omega)$.
- The for-loop runs in $\tilde{\mathcal{O}}(s\mathcal{M}_{q,m}(n))$ operations in $\mathbb{F}_{q^m}$ because the left division in line 4 has complexity $\tilde{\mathcal{O}}(\mathcal{M}_{q,m}(n))$ for each $j = 1, \ldots, s$ according to [11, Section II.D.]. Checking the conditions for a decoding failure is essentially for free.

Note that $\tilde{\mathcal{O}}(s\mathcal{M}_{q,m}(n)) \subseteq \tilde{\mathcal{O}}(sn^{\min(\frac{\omega+1}{2}, 1.635)}) \subseteq \tilde{\mathcal{O}}(sn^{1.635})$. Thus, solving the Gao-like key equation determines the overall complexity of $\tilde{\mathcal{O}}((sn)^\omega)$ operations in $\mathbb{F}_{q^m}$. □

4 A Fast Variant of the Gao-Like Decoder for HILRS Codes

We now present a fast variant of the decoder from Algorithm 1. As we have seen in its complexity analysis in the proof of Theorem 2, the complexity-dominating task is the solution of the Gao-like key equation. Thus, we focus on this problem and obtain a performance gain by reformulating it in terms of minimal approximant bases.

Note that we restrict ourselves to the zero-derivation case in this section, even though the used concepts and algorithms generalize straightforwardly to nonzero derivations. The reason is that the complexity analysis of algorithms involving skew-polynomial operations with nonzero derivations is more involved and was e.g. not conducted for the minimal-approximant-basis algorithm [11, Algorithm 5] that we use for the speedup.

4.1 Minimal Approximant Bases

Let us give some definitions and basic properties of minimal approximant bases. Note that we will only discuss left/row approximant bases and leave out their right/column counterparts, as we are only concerned with these.

Let $\boldsymbol{v} \in \mathbb{Z}^a$ be a shifting vector. Then, the *$\boldsymbol{v}$-shifted row degree* of a vector $\boldsymbol{b} \in \mathbb{F}_{q^m}[x;\theta]^a$ is

$$\operatorname{rdeg}_{\boldsymbol{v}}(\boldsymbol{b}) := \max_{j=1,\ldots,a} \{\deg(b_j + v_j)\}.$$

For $\boldsymbol{b} \in \mathbb{F}_{q^m}[x;\theta]^a \setminus \{\boldsymbol{0}\}$ and $\boldsymbol{v} = (v_1, \ldots, v_b) \in \mathbb{Z}^a$, the *$\boldsymbol{v}$-pivot index* of $\boldsymbol{b}$ is the largest index $i \in \{1, \ldots, a\}$ with $\deg(b_i) + v_i = \operatorname{rdeg}_{\boldsymbol{v}}(\boldsymbol{b})$.

A matrix $\boldsymbol{W} \in \mathbb{F}_{q^m}[x;\theta]^{a\times b}$ with $a \leq b$ is in *$\boldsymbol{v}$-ordered row weak-Popov form* if the $\boldsymbol{v}$-pivot indices of its rows are strictly increasing in the row index.

A vector $\boldsymbol{b} \in \mathbb{F}_{q^m}[x;\theta]^{a}$ is a *left approximant of order* $d \in \mathbb{N}$ of a matrix $\boldsymbol{W} \in \mathbb{F}_{q^m}[x;\theta]^{a\times b}$ if

$$\boldsymbol{b}\boldsymbol{W} \equiv \boldsymbol{0} \bmod_{\mathrm{r}} x^d.$$

A *left $\boldsymbol{v}$-ordered weak-Popov approximant basis of $\boldsymbol{A}$ of order* $d \in \mathbb{N}$ is a full-rank matrix $\boldsymbol{B} \in \mathbb{F}_{q^m}[x;\theta]^{a\times a}$ in $\boldsymbol{v}$-ordered row weak-Popov form whose rows are a basis of all left approximants of $\boldsymbol{A}$ of order d.

4.2 Solving the Gao-Like Key Equation via Minimal Approximant Bases

The Gao-like key equation (7) can also be written as

$$\sigma \cdot f_j = \chi_j \cdot G_j + \sigma \cdot R_j \quad \text{for all } j = 1,\dots,s, \tag{11}$$

where $\chi_j \in \mathbb{F}_{q^m}[x;\theta]$ exists according to the Euclidean algorithm and has degree at most $k+t$ for each $j = 1,\dots,s$. Observe that (11) implies that the vector

$$(\sigma \cdot f_1,\dots,\sigma \cdot f_s, \sigma, \chi_1,\dots,\chi_s) \in \mathbb{F}_{q^m}[x;\theta]^{2s+1}$$

is in the left kernel of the matrix

$$\boldsymbol{W} = \begin{pmatrix} -\boldsymbol{I}_s \\ \boldsymbol{R} \\ \boldsymbol{G} \end{pmatrix} \in \mathbb{F}_{q^m}[x;\theta]^{(2s+1)\times s} \tag{12}$$

where $\boldsymbol{R} := (R_1,\dots,R_s)$ and $\boldsymbol{G} := \mathrm{diag}(G_1,\dots,G_s)$.

The following result based on [11, Lemma 21] is fundamental for reformulating the Gao-like key equation as a minimal-approximant-bases problem.

Lemma 2. *Consider the same setting as in Theorem 2 and let $\boldsymbol{W}$ be defined as in* (12). *Further write*

$$\boldsymbol{\rho} := (\sigma \cdot f_1,\dots,\sigma \cdot f_s,\sigma) \quad \text{and} \quad \boldsymbol{\chi} := (\chi_1,\dots,\chi_s)$$

for simplicity. Further define the shifting vectors $\boldsymbol{w} := (\boldsymbol{0}_s, k-1) \in \mathbb{Z}^{s+1}$ and $\boldsymbol{v} := (\boldsymbol{0}_s, k-1, \boldsymbol{0}_s) \in \mathbb{Z}^{2s+1}$, as well as the degree constraints $D := t_{\mathsf{max}} = \frac{s}{s+1}(n-k)$ and $d := D + n$. Then,

$$(\boldsymbol{\rho} \mid \boldsymbol{\chi}) \cdot \boldsymbol{W} = \boldsymbol{0} \quad \textit{and} \quad \mathrm{rdeg}_{\boldsymbol{w}}(\boldsymbol{\rho}) < D \tag{13}$$

if and only if

$$(\boldsymbol{\rho} \mid \boldsymbol{\chi}) \cdot \boldsymbol{W} \equiv \boldsymbol{0} \bmod_{\mathrm{r}} x^d \quad \textit{and} \quad \mathrm{rdeg}_{\boldsymbol{v}}(\boldsymbol{\rho} \mid \boldsymbol{\chi}) < D. \tag{14}$$

Proof. We start with showing that (13) implies (14). The left-hand side of (14) clearly follows from (13) and it remains to show that $\deg(\chi_j) < D$ holds for all $j = 1, \ldots, s$. With (11), we get

$$\begin{aligned}\deg(\chi_j) &\leq \max\{\deg(\sigma \cdot f_j), \deg(\sigma \cdot R_j)\} - \deg(G_j)\\ &\leq \max\{t + k - 1, t + n - 1\} - n < t \leq t_{\mathsf{max}} = D.\end{aligned}$$

For the other implication, note that the right-hand side of (14) directly implies the right-hand side of (13). In order to see that the left-hand side of (13) holds, we show that all entries of the vector $(\boldsymbol{\rho} \mid \boldsymbol{\chi}) \cdot \boldsymbol{W}$ have degree less than d. With the help of the right-hand side of (14) and (11), we obtain:

- $\deg(\sigma \cdot f_j) < D < d$,
- $\deg(\sigma \cdot R_j) \leq \deg(\sigma) + \deg(R_j) \leq t + n - 1 = D + n - 1 < d$,
- $\deg(\chi_j \cdot G_j) < t + n = D + n = d$.

□

Hence, we can solve the Gao-like key Eq. (7) by computing a left $\boldsymbol{v}$-ordered weak-Popov approximant basis $\boldsymbol{B}$ of $\boldsymbol{W}$. This can be accomplished by [11, Algorithm 5] requiring $\tilde{O}(\mathcal{M}(n)) \subseteq \tilde{O}(n^{\min\{\frac{\omega+1}{2}, 1.635\}}) \subseteq \tilde{O}(n^{1.635})$ operations in $\mathbb{F}_{q^m}$.

We then obtain candidates p_j for the products $\sigma \cdot f_j$ for each $j = 1, \ldots, s$ and a candidate $\tilde{\sigma}$ for the σ by choosing the row $\boldsymbol{b}_{\mathsf{min}}$ of $\boldsymbol{B}$ having minimal $\boldsymbol{v}$-weighted degree. This choice makes sure to satisfy the degree constraint in (14) to get a proper solution as described in Lemma 2. The subroutine for solving the Gao-like key equation via the presented minimal-approximant-bases approach is summarized in Algorithm 2.

Algorithm 2: Subroutine $\mathsf{solveKEviaMAB}(\cdot)$ for Solving the Gao-like Key Equation via a Minimal Approximant Basis

---

Input : $R_1, \ldots, R_s, G_1, \ldots, G_s, n, k, s$
Output : $p_1, \ldots, p_s, \tilde{\sigma}$

1 $\boldsymbol{v} := (\boldsymbol{0}_s, k-1, \boldsymbol{0}_s)$
2 $D := \frac{s}{s+1}(n-k)$ and $d := D + n$
3 $\boldsymbol{W} := \begin{pmatrix} -\boldsymbol{I}_s \\ \boldsymbol{R} \\ \boldsymbol{G} \end{pmatrix} \in \mathbb{F}_{q^m}[x;\theta]^{(2s+1)\times s}$ with $\boldsymbol{R} := (R_1, \ldots, R_s)$ and
$\boldsymbol{G} := \operatorname{diag}(G_1, \ldots, G_s)$
`/* left` $\boldsymbol{v}$`-ordered weak Popov approximant basis of` $\boldsymbol{W}$ `of order` d `*/`
4 $\boldsymbol{B} := \mathsf{LeftSkewPMBasis}(d, \boldsymbol{W}, \boldsymbol{v}) \in \mathbb{F}_{q^m}[x;\theta]^{(2s+1)\times(2s+1)}$
5 Define $\boldsymbol{b}_{\mathsf{min}} = (b_{\mathsf{min},1}, \ldots, b_{\mathsf{min},2s+1})$ as the minimal row of $\boldsymbol{B}$ with respect to the $\boldsymbol{v}$-weighted degree
6 **return** $b_{\mathsf{min},1}, \ldots, b_{\mathsf{min},s}, b_{\mathsf{min},s+1}$

Theorem 3. *Algorithm 2 solves the Gao-like key Eq. (7) in* $\tilde{O}(s^{\omega}n^{1.635}) \subseteq \tilde{O}(s^{2.373}n^{1.635})$ $\mathbb{F}_{q^m}$*-operations.*

Proof. The complexity of Algorithm 2 is dominated by finding a minimal approximant basis in line 4. This can be achieved using [11, Algorithm 5] whose complexity is $\tilde{O}(s^{\omega}n^{1.635}) \subseteq \tilde{O}(s^{2.373}n^{1.635})$ [11, Theorem 11].

This directly implies the following complexity improvement for Theorem 2:

Corollary 1. *When the Gao-like key Eq. (7) is solved by Algorithm 2, the complexity of the Gao-like decoder from Algorithm 1 decreases to* $\tilde{O}(s^{\omega}n^{1.635}) \subseteq \tilde{O}(s^{2.373}n^{1.635})$ *operations in* $\mathbb{F}_{q^m}$.

With Corollary 1, the Gao-like decoder is the fastest known decoder for HILRS codes in the sum-rank metric as well as for horizontally interleaved Gabidulin codes in the rank metric. Its complexity is essentially subquadratic in the component-code length n, as the interleaving order s is usually much smaller than the code length n. Remark in particular that the gain in the error-correcting capacity increases fast for increasing s, as $\frac{s}{s+1}$ quickly tends to one.

5 Conclusion

We studied HILRS codes and their fast decoding which has promising potential applications in code-based cryptography. As a starting point, we presented a Gao-like decoder that features probabilistic unique decoding for an error of sum-rank weight at most $\frac{s}{s+1}(n-k)$, where s is the interleaving order, and n and k are the length and the dimension of the component codes. We gave a bound on the failure probability and achieved a complexity of $\tilde{O}((sn)^{2.373})$ operations in $\mathbb{F}_{q^m}$ by solving the Gao-like key equation conventionally via Gaussian elimination.

Techniques from the area of minimal approximant bases allowed us to speed up the decoder significantly and obtain a complexity of $\tilde{O}(s^{2.373}n^{1.635})$ operations in $\mathbb{F}_{q^m}$. Under the reasonable assumption that the interleaving order s is small compared to the component-code length n, this is subquadratic. Overall, this results in the fastest known decoders for both HILRS codes in the sum-rank metric and for horizontally interleaved Gabidulin codes in the rank metric.

Further work can include the generalization of the presented decoder to the error-erasure case. Next to errors, this error model includes row and column erasures, for which either the row space or the column space is known. Moreover, other techniques could give bounds on the failure probability for nonzero derivations or yield tighter ones for the zero-derivation setting.

References

1. Aguilar Melchor, C., et al.: Hamming Quasi-Cyclic (HQC) (2023). http://pqc-hqc.org/download.php?file=hqc-specification_2023-04-30.pdf

2. Aguilar-Melchor, C., Aragon, N., Dyseryn, V., Gaborit, P., Zémor, G.: LRPC codes with multiple syndromes: near ideal-size KEMs without ideals. In: Cheon, J.H., Johansson, T. (eds.) PQCrypto 2022. LNCS, vol. 13512, pp. 45–68. Springer, Cham (2022). https://doi.org/10.1007/978-3-031-17234-2_3
3. Alagic, G., et al.: Status report on the third round of the NIST post-quantum cryptography standardization process (2022). https://doi.org/10.6028/NIST.IR.8413-upd1
4. Aragon, N., et al.: BIKE: bit flipping key encapsulation (2022). https://bikesuite.org/files/v5.0/BIKE_Spec.2022.10.10.1.pdf
5. Aragon, N., Blazy, O., Gaborit, P., Hauteville, A., Zémor, G.: Durandal: a rank metric based signature scheme. In: Ishai, Y., Rijmen, V. (eds.) EUROCRYPT 2019. LNCS, vol. 11478, pp. 728–758. Springer, Cham (2019). https://doi.org/10.1007/978-3-030-17659-4_25
6. Aragon, N., Dyseryn, V., Gaborit, P., Loidreau, P., Renner, J., Wachter-Zeh, A.: LowMS: a new rank metric code-based KEM without ideal structure. Cryptology ePrint Archive, Paper 2022/1596 (2022)
7. Aragon, N., Gaborit, P., Hauteville, A., Tillich, J.P.: A new algorithm for solving the rank syndrome decoding problem. In: IEEE International Symposium on Information Theory (ISIT), pp. 2421–2425 (2018)
8. Avanzi, R., et al.: CRYSTALS-Kyber: algorithm specifications and supporting documentation (Version 3.02) (2021). https://pq-crystals.org/kyber/data/kyber-specification-round3-20210804.pdf
9. Azarderakhsh, R., et al.: Supersingular isogeny key encapsulation (2022). https://sike.org/files/SIDH-spec.pdf
10. Bardet, M., Briaud, P.: An algebraic approach to the rank support learning problem. In: Cheon, J.H., Tillich, J.-P. (eds.) PQCrypto 2021 2021. LNCS, vol. 12841, pp. 442–462. Springer, Cham (2021). https://doi.org/10.1007/978-3-030-81293-5_23
11. Bartz, H., Jerkovits, T., Puchinger, S., Rosenkilde, J.: Fast decoding of codes in the rank, subspace, and sum-rank metric. IEEE Trans. Inf. Theory **67**(8), 5026–5050 (2021)
12. Bartz, H., Puchinger, S.: Decoding of interleaved linearized Reed-Solomon codes with applications to network coding. In: IEEE International Symposium on Information Theory (ISIT), pp. 160–165 (2021)
13. Bartz, H., Puchinger, S.: Fast decoding of interleaved linearized Reed-Solomon codes and variants. IEEE Trans. Inf. Theory (2023, submitted). https://arxiv.org/abs/2201.01339v3
14. Bernstein, D.J., et al.: Classic McEliece: conservative code-based cryptography: cryptosystem specification (2022). https://classic.mceliece.org/mceliece-spec-20221023.pdf
15. Boucher, D.: An algorithm for decoding skew Reed-Solomon codes with respect to the skew metric. Des. Codes Crypt. **88**(9), 1991–2005 (2020)
16. Caruso, X.: Residues of skew rational functions and linearized Goppa codes. arXiv preprint arXiv:1908.08430v1 (2019)
17. Castryck, W., Decru, T.: An efficient key recovery attack on SIDH. Cryptology ePrint Archive, Paper 2022/975 (2022)
18. Gabidulin, E.M., Paramonov, A.V., Tretjakov, O.V.: Ideals over a non-commutative ring and their application in cryptology. In: Davies, D.W. (ed.) EUROCRYPT 1991. LNCS, vol. 547, pp. 482–489. Springer, Heidelberg (1991). https://doi.org/10.1007/3-540-46416-6_41

19. Gabidulin, E.M., Rashwan, H., Honary, B.: On improving security of GPT cryptosystems. In: IEEE International Symposium on Information Theory, pp. 1110–1114 (2009)
20. Gao, S.: A new algorithm for decoding Reed-Solomon codes. In: Bhargava, V.K., Poor, H.V., Tarokh, V., Yoon, S. (eds.) Communications, Information and Network Security. The Springer International Series in Engineering and Computer Science, vol. 712, pp. 55–68. Springer, Boston (2003). https://doi.org/10.1007/978-1-4757-3789-9_5
21. Hörmann, F., Bartz, H., Horlemann, A.L.: Distinguishing and recovering generalized linearized Reed-Solomon codes. In: Deneuville, J.C. (ed.) CBCrypto 2022. LNCS, vol. 13839, pp. 1–20. Springer, Cham (2023). https://doi.org/10.1007/978-3-031-29689-5_1
22. Hörmann, F., Bartz, H., Puchinger, S.: Error-erasure decoding of linearized Reed-Solomon codes in the sum-rank metric. In: IEEE International Symposium on Information Theory (ISIT), pp. 7–12 (2022)
23. Hörmann, F., Bartz, H., Puchinger, S.: Syndrome-based error-erasure decoding of interleaved linearized Reed-Solomon codes. IEEE Trans. Inf. Theory (2023, submitted)
24. Jerkovits, T., Hörmann, F., Bartz, H.: On decoding high-order interleaved sum-rank-metric codes. In: Deneuville, J.C. (ed.) CBCrypto 2022. LNCS, vol. 13839, pp. 90–109. Springer, Cham (2023). https://doi.org/10.1007/978-3-031-29689-5_6
25. Lam, T.Y., Leroy, A.: Vandermonde and Wronskian matrices over division rings. J. Algebra **119**(2), 308–336 (1988)
26. Le Gall, F.: Powers of tensors and fast matrix multiplication. In: Proceedings of the 39th International Symposium on Symbolic and Algebraic Computation, pp. 296–303 (2014)
27. Loidreau, P.: An evolution of GPT cryptosystem. In: International Workshop on Algebraic and Combinatorial Coding Theory (ACCT) (2016)
28. Loidreau, P.: Designing a rank metric based McEliece cryptosystem. In: Sendrier, N. (ed.) PQCrypto 2010. LNCS, vol. 6061, pp. 142–152. Springer, Heidelberg (2010). https://doi.org/10.1007/978-3-642-12929-2_11
29. Martínez-Peñas, U.: Skew and linearized Reed-Solomon codes and maximum sum rank distance codes over any division ring. J. Algebra **504**, 587–612 (2018)
30. Martínez-Peñas, U., Kschischang, F.R.: Reliable and secure multishot network coding using linearized Reed-Solomon codes. IEEE Trans. Inf. Theory **65**(8), 4785–4803 (2019)
31. Matsaglia, G., Styan, G.P.H.: Equalities and inequalities for ranks of matrices. Linear Multilinear Algebra **2**(3), 269–292 (1974)
32. McEliece, R.J.: A public-key cryptosystem based on algebraic coding theory. Deep Space Netw. Progr. Rep. **42–44**, 114–116 (1978)
33. Metzner, J., Kapturowski, E.: A general decoding technique applicable to replicated file disagreement location and concatenated code decoding. IEEE Trans. Inf. Theory **36**(4), 911–917 (1990)
34. Ore, O.: On a special class of polynomials. Trans. Am. Math. Soc. **35**(3), 559–584 (1933)
35. Ore, O.: Theory of non-commutative polynomials. Ann. Math. 480–508 (1933)
36. Puchinger, S., Rosenkilde né Nielsen, J., Li, W., Sidorenko, V.: Row reduction applied to decoding of rank-metric and subspace codes. Des. Codes Crypt. **82**(1–2), 389–409 (2017)
37. Puchinger, S., Renner, J., Rosenkilde, J.: Generic decoding in the sum-rank metric. In: IEEE International Symposium on Information Theory (ISIT), pp. 54–59 (2020)

38. Puchinger, S., Renner, J., Wachter-Zeh, A.: Decoding high-order interleaved rank-metric codes. arXiv preprint arXiv:1904.08774 (2019)
39. Renner, J., Jerkovits, T., Bartz, H.: Efficient decoding of interleaved low-rank parity-check codes. In: 2019 XVI International Symposium "Problems of Redundancy in Information and Control Systems" (REDUNDANCY), pp. 121–126 (2019)
40. Renner, J., Puchinger, S., Wachter-Zeh, A.: Decoding high-order interleaved rank-metric codes. In: IEEE International Symposium on Information Theory (ISIT), pp. 19–24 (2021)
41. Sidorenko, V., Bossert, M.: Decoding interleaved Gabidulin codes and multisequence linearized shift-register synthesis. In: IEEE International Symposium on Information Theory, pp. 1148–1152 (2010)
42. Sidorenko, V., Jiang, L., Bossert, M.: Skew-feedback shift-register synthesis and decoding interleaved Gabidulin codes. IEEE Trans. Inf. Theory **57**(2), 621–632 (2011)
43. Stein, W.A., et al.: Sage Mathematics Software (Version 9.6). The Sage Development Team (2022). http://www.sagemath.org
44. Storjohann, A.: Algorithms for matrix canonical forms. Ph.D. thesis (2000)
45. Wachter-Zeh, A.: Decoding of block and convolutional codes in rank metric. Ph.D. thesis, Ulm University and University of Rennes 1, Ulm, Germany and Rennes, France (2013)
46. Wachter-Zeh, A., Afanassiev, V., Sidorenko, V.: Fast decoding of Gabidulin codes. Des. Codes Crypt. **66**(1), 57–73 (2013)

Theoretical Analysis of Decoding Failure Rate of Non–binary QC–MDPC Codes

Kirill Vedenev(✉) and Yury Kosolapov

Southern Federal University, Rostov-on-Don, Russia
vedenevk@gmail.com

Abstract. In this paper, we study the decoding failure rate (DFR) of non-binary QC-MDPC codes using theoretical tools, extending the results of previous binary QC-MDPC code studies. The theoretical estimates of the DFR are particularly significant for cryptographic applications of QC-MDPC codes. Specifically, in the binary case, it is established that exploiting decoding failures makes it possible to recover the secret key of a QC-MDPC cryptosystem. This implies that to attain the desired security level against adversaries in the CCA2 model, the decoding failure rate must be strictly upper-bounded to be negligibly small. In this paper, we observe that this attack can also be extended to the non–binary case as well, which underscores the importance of DFR estimation. Consequently, we study the guaranteed error–correction capability of non–binary QC–MDPC codes under one–step majority logic (OSML) decoder and provide a theoretical analysis of the 1–iteration parallel symbol flipping decoder and its combination with OSML decoder. Utilizing these results, we estimate the potential public-key sizes for QC-MDPC cryptosystems over $\mathbb{F}_4$ for various security levels. We find that there is no advantage in reducing key sizes when compared to the binary case.

Keywords: code–based cryptography · non–binary MDPC codes · symbol flipping · decoding failure rate

1 Introduction

With the advent of quantum computers, many traditional public–key cryptosystems based on number–theoretic or elliptic curves primitives are to become vulnerable to attacks using them [14,42]. So, there is a strong need in developing *post-quantum* cryptographic protocols that will remain secure against adversaries equipped with quantum computers. One of the most prominent and well-established approach to post-quantum cryptography is cryptography based on error-correcting codes.

The first code–based cryptosystem was proposed in 1978 by R. McEliece [31]. The main idea of the McEliece cryptosystem is to mask a generator matrix of a fast–decodable code by permuting its columns and multiplying by a scrambling matrix on the left. The encryption is performed by encoding a message using the

A. Esser and P. Santini (Eds.): CBCrypto 2023, LNCS 14311, pp. 35–55, 2023.
https://doi.org/10.1007/978-3-031-46495-9_3

public generator matrix and adding an error. So, the security against *message–recovery attacks* is based on NP–hard syndrome decoding problem [13]. The original proposal of R. McEliece was based on binary Goppa codes, so the security against *key–recovery attack* relies on hardness of the problem of distinguishing permuted Goppa codes. It is worth mentioning that the original McEliece cryptosystem with several improvements is one of three Round 4 competitors in NIST-PQC [1]. Despite many advantages, the main drawback of McEliece cryptosystem is large public–key size. There were many attempts to overcome this by replacing Goppa codes with more efficient ones. The notable examples are Generalized Reed–Solomon codes [35], Reed–Muller codes [43], algebraic geometry codes [29], concatenated codes [40], rank–metric Gabidulin codes [22]. However, most of this modifications were proven unsecure [15,17,32,38,40,44]. In addition, several modifications of protocol itself were proposed to avoid key–recovery attacks against McEliece–like cryptosystems based on efficient algebraic codes (e.g. [7,12,28,48]), however most of them were also successfully cryptanalyzed [16,18,19,30,47,50].

One of the most efficient approaches to reducing public-key size was proposed by P. Gaborit [23] and is based on using *quasi-cyclic codes (QC-codes)*. A code C of length $n = n'l$ is said to be quasi-cyclic of order n' and index l if its permutation automorphism group $\mathrm{PAut}(C)$ has a cyclic subgroup of order n' that acts freely on coordinates. The quasi-cyclic structure implies the existence of generator and parity-check matrices of C that admit a *block-circulant* representation, i.e.

$$\begin{pmatrix} \mathrm{rot}(h_{1,1}) \dots \mathrm{rot}(h_{1,l}) \\ \vdots \quad \ddots \quad \vdots \\ \mathrm{rot}(h_{s,1}) \dots \mathrm{rot}(h_{s,l}) \end{pmatrix}, \quad \mathrm{rot}(a_1, a_2, \dots, a_{n'}) = \begin{pmatrix} a_1 & a_2 & \dots & a_{n'} \\ a_n & a_1 & \dots & a_{n'-1} \\ \vdots & \vdots & \ddots & \vdots \\ a_2 & a_3 & \dots & a_1 \end{pmatrix}. \tag{1}$$

This representation allows storing only the first row of each circulant block $\mathrm{rot}(h_{i,j})$, thereby reducing storage and communication costs. Therefore, the public key sizes of code-based encryption protocols that preserve quasi-cyclic structure can be significantly reduced. Note that many encryption protocols based on *algebraic* QC–codes (e.g. [12,23]) have been successfully attacked [20,36]. However, protocols based on *random quasi–cyclic moderate density parity–check (QC–MDPC) codes* [33], which have no algebraic structure except being quasi–cyclic, are still considered secure and efficient.

The concept of *moderate-density parity-check (MDPC) codes* extends the idea of *low-density parity-check codes (LDPC codes)* initially introduced by R. Gallager [24]. In Gallager's seminal work [24], it was shown that efficient decoding of binary codes with a parity-check matrix containing a very small constant number of ones in each row is feasible using iterative algorithms such as bit-flipping and belief propagation, provided certain conditions are met (no two rows have two or more ones in the same positions). In 2000, C. Monico et al. [34] considered replacing Goppa codes in the McEliece cryptosystem with LDPC codes and pointed out that these codes can be easily distinguished due to the existence of very low–weight codewords in the dual code. The application of quasi-cyclic LDPC codes in constructing code-based cryptosystems was initially proposed in [11] and further developed in [8,10]. To mitigate key-recovery attacks based on searching for low-weight dual codewords, it was suggested to replace the permutation matrix in the protocol with a sparse non–singular matrix of a specific form. However, this approach was found to introduce serious vulnerabilities [2,36]. An alternative method to prevent key–recovery based on the search for low–weight codewords was proposed in [33], where it was suggested to use random QC–MDPC codes instead of LDPC. The difference between MDPC and LDPC codes lies in the slightly higher weight of the rows in the parity-check matrices, i.e., which is of order $O(\sqrt{n})$ for MDPC codes and $O(1)$ for LDPC.

We denote the finite field of size q as $\mathbb{F}_q$. For a vector $v \in \mathbb{F}_q^n$, the notation $\mathrm{supp}(v) = \{i \in [\![1, n]\!] \mid v_i \neq 0\}$ is used to represent the set of indices corresponding to the positions where v is nonzero. Here, $[\![a, b]\!] = \{a, a+1, \ldots, b\}$ denotes set of all integers between a and b. The Hamming weight of vector v, denoted as $\mathrm{wt}(v)$, is defined as the number of nonzero positions in v. A linear code $C \in \mathbb{F}_q^n$ of length n and dimension k is refereed as $[n, k]_q$–code. A generic description of a QC-MDPC cryptosystem [33] in the Niedderiter form [35] is as follows:

- **Key generation** The secret key is the parity-check matrix H of a random QC-MDPC $[n = ln', (l-1)n']_q$-code, represented as

$$H = \left(H_1 \mid H_2 \mid \ldots \mid H_{l-1} \mid H_l\right). \tag{2}$$

  The matrix H consists of circulant $(n' \times n')$–blocks H_i, where each H_i has a row weight of γ. Note that n' is usually chosen to be a prime number p. The public key is the systematic form of H, i.e.

$$\tilde{H} = H_1^{-1}H = \left(I_{n'} \mid H_1^{-1}H_2 \mid \ldots \mid H_1^{-1}H_{l},\right)$$

  which can be represented by the first rows of $H_1^{-1}H_i$, where $i \in [\![2, l]\!]$, since the product of circulant matrices is also a circulant matrix.
- **Encryption** The plaintext is an error vector $e \in \mathbb{F}_q^n$ of weight t, and the ciphertext is its syndrome $\tilde{s} = \tilde{H}e^{\mathsf{T}}$.
- **Decryption** To decrypt, the private syndrome $s = He^{\mathsf{T}} = H_1\tilde{s}^{\mathsf{T}}$ is computed and used as input for the MDPC decoder (bit-flipping or symbol flipping).

Note that in NIST-PQC, the QC–MDPC approach is represented by *BIKE (bit-flipping key encapsulation)* protocol [3].

Due to probabilistic nature of decoding of LDPC and MDPC codes, there is a non–zero probability of decryption failure. In [26] it was shown that decryption failures can be used to recover the secret key in binary case. Hence in order to achieve *indistinguishability against chosen ciphertext attacks*, where an adversary has an access to a decryption oracle (IND–CCA2 *security*), the *decoding failure rate (DFR)* has to be negligibly small, i.e. of order $2^{-\lambda}$, where λ is a security level. In [46], an experimental–based extrapolation framework for estimating DFR has been proposed. In this approach, the DFR curve is assumed to be concave, so

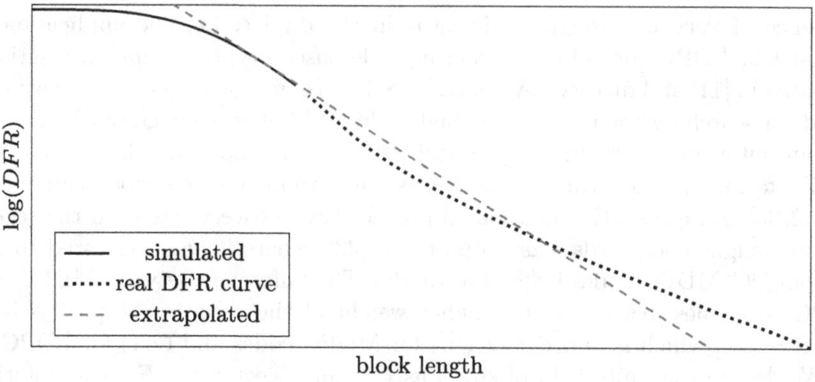

Fig. 1. Approximate illustration of a situation where the use of extrapolation may lead to an incorrect estimation of DFR due to the presence of an error floor.

estimates for high DFR ($> 10^{-9}$) can be obtained via numerical simulations and then extrapolated to low DFRs providing an upper bound. However, it is known that LDPC and MDPC codes exhibit error floor phenomenon, resulting in violation of concavity assumption (see e.g. [4,6]). Hence DFR estimates obtained by extrapolation could possibly be overly optimistic (see Fig. 1). Another approach is to estimate DFR using only theoretical tools. In [45] J. P. Tillich studied guaranteed error–correction performance of binary QC–MDPC codes under one–step majority logic decoder (OSML). In addition, in [45] the DFR of two–iteraion decoder is studied under some reasonable assumptions, i.e. the probability that one iteration of parallel bit–flipping decoder reduces error weight enough to be corrected by OSML decoder is computed. In [39], the estimate of the number of errors correctable by OSML decoder was improved. Under the same assumptions as in [45], the worst–case plausibility analysis of one and two iteration randomized serial bit–flipping decoder was performed in [5]. In addition, in [5] a combination of one iteration of randomized serial bit-flipping and OSML was studied, and recommended design parameters for IND–CCA2 secure QC-MDPC cryptosystems were given.

In this paper, we study DFR of *non–binary* QC–MDPC codes using theoretical tools. Namely, we extend the results of [39,45] to the non–binary case, i.e. we show that error–correcting performance of OSML decoder can also be estimated using similar methods of [39,45]. In addition, we propose a parallel symbol flipping decoder. Under the same assumptions used in [5,45], we give theoretical estimates of DFR for the parallel symbol–flipping decoder and its combination with the OSML decoder. We also note that the extension of the randomized serial approach, as considered in [5], in the non-binary case seems to yield unreliable results due to a observed discrepancy between the theoretical estimates and the worst-case simulations. Hence this approach is not included in this paper. In addition, we experimentally demonstrate that slightly modified attack of [26] can also recover secret key in non–binary case. Employing the obtained results, recommended parameters and corresponding key sizes for IND-CCA2–secure QC–MDPC cryptosystems over $\mathbb{F}_q$ for various security levels are computed.

The paper is organized as follows. In Sect. 2, we present the basic principles of decoding non-binary QC-MDPC codes and study the guaranteed error-correction capability of the one-step majority logic decoder in an assumption-free setting. In Sect. 3, we provide a plausibility analysis of error counters distribution and flipping probability in the non-binary case. Subsequently, we propose a 1-iteration parallel symbol flipping decoder and theoretically estimate the probability of reducing the error weight to a certain value, allowing for further decoding by the OSML decoder. We also provide experimental validation of the theoretical model. Finally, in Sect. 4, we consider the reaction attack against non–binary QC–MDPC cryptosystems and find potential cryptosystem parameters and corresponding public-key sizes.

2 Analysis of Guaranteed Error–correction Capability of Non–binary QC-MDPC Codes

Recall that a code C with a parity–check matrix $H \in \mathbb{F}_q^{m\times n}$ is said to be a *moderate–density parity–check (MDPC) code* if each row of the $H = (h_{i,j})$ is of weight $O(\sqrt{n})$. In addition, C is said to be (γ, δ)–regular if the weight of each column of H is γ and the weight of each row is δ. Unless otherwise specified, we will focus exclusively on regular MDPC codes.

Let $z = c + e \in \mathbb{F}_q^n$, where $c \in C$ and $\mathrm{wt}(e) \leq t$, be a noisy codeword. By $s = Hz^\mathsf{T} = He^\mathsf{T}$ we denote the syndrome of e. One can easily note that since i–th position of s is computed as

$$s_i = \langle h_i, e\rangle = \sum_{\omega\in\mathsf{supp}\,(h_i)} h_{i,\omega}e_\omega.$$

Hence, by selecting γ row indices $i_1, i_2, \ldots, i_\gamma$ for which $h_{i_1,j}, \ldots, h_{i_1,j}$ are non-zero, we obtain the following γ equalities:

$$\begin{cases} s_{i_1}h_{i_1,j}^{-1} = e_j + h_{i_1,j}^{-1}\left(\sum_{\omega\in\mathsf{supp}(h_{i_1})\setminus\{j\}} h_{i_1,\omega}e_\omega\right), \\ \ldots \\ s_{i_\gamma}h_{i_\gamma,j}^{-1} = e_j + h_{i_\gamma,j}^{-1}\left(\sum_{\omega\in\mathsf{supp}(h_{i_\gamma})\setminus\{j\}} h_{i_\gamma,\omega}e_\omega\right). \end{cases} \tag{3}$$

Since C is an MDPC code, the rows h_i of H are sparse. Considering sparsity of e, it follows that $s_ih_{i,j}^{-1}$ equals e_j with high probability. Hence it is possible to use the values $s_ih_{i,j}^{-1}$ for estimating e.

Let $\mathbb{F}_q = \{\alpha_0 = 0, \alpha_1 = 1, \alpha_2, \ldots, \alpha_{q-1}\}$ be a enumeration of elements of $\mathbb{F}_q$. Let us define

$$\sigma_{j,i} = \left|\{w \mid h_{w,j} \neq 0 \text{ and } s_ih_{w,j}^{-1} = \alpha_i\}\right| \tag{4}$$

as the number of rows h_w containing the position j in $\mathrm{supp}(h_w)$ and $s_ih_{w,j}^{-1} = \alpha_i$. The values of $\sigma_{j,i}$ will be referred to as *error counters* in position j. Clearly, $\sigma_{j,i}$

indicates the likelihood that the error value e_j in position j is equal to α_i. In particular, a higher value of $\sigma_{j,0}$ implies that position j is less likely to be corrupted, while higher values of $\sigma_{j,i}$, $i \neq 0$, indicate a greater likelihood that $e_j = \alpha_i \neq 0$.

Therefore, several decoding strategies are possible. For instance, it is possible to choose an information set I of k positions with highest $\sigma_{j,0}$, indicating that these positions less likely to be erroneous, and then use this I for information set decoding *(ordered statistics decoding* [21] *and statistical decoding* [37]*)*.

Another straightforward decoding algorithm that uses counters is as follows:

1. compute the syndrome s and the counters $\sigma_{j,i}$ for all $j \in [\![1, n]\!]$ and $i \in [\![0, q-1]\!]$;
2. update the position j of the received word z having the highest value of $\sigma_j^* - \sigma_{j,0}$, where
$$\sigma_j^* = \max_{i \in [\![1,q-1]\!]} \sigma_{j,i}, \tag{5}$$
to the new value $z_j - \alpha_{i^*}$, where $i^* = \operatorname{argmax}_{i \in [\![1,q-1]\!]} \sigma_{j,i}$;
3. repeat from step 1 until either $s = 0$ or maximum number of iterations is reached.

Remark 1. One can easily note that the syndrome weight after step 2 is decreased by $\sigma_j^* - \sigma_{j,0}$. Therefore, the error position and error value in step 2 are chosen to decrease the syndrome weight the most. In this formulation the decoding approach described above was proposed in [9] as a generalization of Gallager's bit–flipping. In the binary case, the Gallagher's decoder is also a greedy algorithm that reduces the syndrome weight the most in each step .

2.1 One–Step Majority Logic Decoding

In this subsection, we study guaranteed decoding performance of regular MDPC codes under the OSML decoder (Algorithm 1) which can be considered as single iteration version of parallel symbol flipping.

Algorithm 1: OSML

```
Input: syndrome s = He^T
Output: estimated error ẽ
ẽ ← 0^n,  s ← Hz^T;
for j ← 1 to n do
    using (4), (5) compute σ_j = (σ_{j,0}, ..., σ_{j,q-1}) and σ*_j ;
    if σ*_j ≥ th_j then
        l ← argmax_{i∈[[1,q-1]]} σ_{j,i};
        ẽ_j ← ẽ_j + α_l;
end
return ẽ
```

Remark 2. Note that in the decoder description, instead of recovering the corrected codeword $c \in C$ from the noisy codeword $z = c + e$ by iteratively subtracting the estimated error from z, we employ an equivalent formulation where we iteratively find the estimated error $\tilde{e}$ itself.

Let $X \in \mathbb{F}_q^{m \times n}$ be an $(m \times n)$–matrix, and let $I \subset [\![1, m]\!]$ and $J \subset [\![1, n]\!]$ be sets of row and column indices, respectively. We denote the matrix composed of the elements of X with indices $(i, j) \in I \times J$ as $X_{I,J} = (x_{i,j})_{i \in I, j \in J}$. For convenience, we use the notations $X_{:,J}$ and $X_{I,:}$ to represent $X_{[\![1,m]\!],J}$ and $X_{I,[\![1,n]\!]}$, respectively.

Proposition 1. *Let $H = (h_{i,j}) \in \mathbb{F}_q^{m \times n}$ be a parity–check matrix of a MDPC code, and let $e \in \mathbb{F}_q^n$ be an error of weight t. Define $H^{(j)}$ as the matrix consisting of rows from the set*

$$\left\{h_{i,j}^{-1} \cdot \left(H_{i,\, [\![1,n]\!] \setminus \{j\}}\right) \mid i \in [\![1, m]\!],\ h_{i,j} \neq 0\right\}.$$

Let

$$a_l = \mathrm{wt}(H_{:,l}^{(j)}), \quad \mu(s) = \sum_{\substack{\omega \in \text{indicies of } s \text{ largest} \\ \text{values of } a_l}} a_\omega,$$

If $e_j = \alpha_i$, then $\sigma_{j,i}$ can be lower bounded as follows

$$\sigma_{j,i} \geq \begin{cases} \gamma - \mu(t), & e_j = \alpha_i = 0, \\ \gamma - \mu(t-1), & e_j = \alpha_i \neq 0. \end{cases}$$

Proof. Using (3), we obtain that $\sigma_{j,i}$ denotes the frequency of occurrence of α_i in the vector

$$v = \begin{pmatrix} s_{i_1} h_{i_1,j}^{-1} \\ \vdots \\ s_{i_\gamma} h_{i_\gamma,j}^{-1} \end{pmatrix} = \begin{pmatrix} e_j \\ \vdots \\ e_j \end{pmatrix} + \underbrace{H^{(j)} e'^{\mathsf{T}}}_{v'}, \quad e' = e_{[\![1,n]\!] \setminus \{j\}}.$$

Hence if $e_j = \alpha_i$ then $\sigma_{j,i} = \gamma - \mathrm{wt}(v')$. Since v' is a linear combination of $\mathrm{wt}(e')$ columns of $H^{(j)}$, its weight can be upper bounded by

$$\mathrm{wt}(v') \leq \mu(\mathrm{wt}(e')) = \begin{cases} \mu(t), & e_j = 0, \\ \mu(t-1), & e_j \neq 0. \end{cases}$$

This concludes the proof of the proposition.

Remark 3. Note that the weight $\mathrm{wt}(H_{:,l}^{(j)})$ of l–th column $H_{:,l}^{(j)}$ of $H^{(j)}$ equals

$$\left|\mathrm{supp}\,(H_{:,l}) \cap \mathrm{supp}\,(H_{:,j})\right|.$$

Corollary 1. *Let $\mathrm{wt}(e) \leq t$. If $\mu(t) < th_j \leq \gamma - \mu(t-1)$, then the OSML decoder correctly estimates the j–th position of e.*

Proof. If $e_j = 0$, then $\sigma_{j,0} \geq \gamma - \mu(t)$ and hence $\sigma_j^* \leq \gamma - \sigma_{j,0} \leq \mu(t)$. It follows that setting $th_j > \mu(t)$ in Algorithm 1 will ensure that no non–erroneous position will be corrupted.

If $e_j = \alpha_i \neq 0$, then $\sigma_{j,i} \geq \gamma - \mu(t-1)$. Since $\mu(t) < \gamma - \mu(t-1)$ and $\mu(t) \geq \mu(t-1)$, it follows that $\mu(t-1) < \gamma/2$ and thereby $\sigma_{j,i} \geq \gamma - \mu(t-1) > \gamma/2$. This implies that a clear majority of equalities in (3) vote for α_i and hence $\sigma_j^* = \sigma_{j,i}$ (see (5)). Therefore, setting $th_j < \gamma - \mu(t-1)$ will ensure that error value in a erroneous position will be estimated correctly.

Corollary 2. *The guaranteed error–correction capability of OSML decoder is t if for all $j \in [\![1, n]\!]$ it is possible to choose th_j according to Corollary 1.*

Note that OSML is a very simple yet effective decoder that is capable of correcting low–weight error patterns. However, it is particularly useful as a second decoding iteration because it does not rely on probabilistic assumptions. It can effectively decode errors of a certain weight that remain after previous iterations, even if they have a harder–to–decode structure that would make plausibility analysis based on probabilistic assumptions irrelevant.

3 Plausibility Analysis of 1–iteration Parallel Symbol Flipping Decoder

In this section, we provide an analysis of the single-iteration parallel symbol flipping algorithm. Namely, following the approach of [45], we estimate the probability of correcting an error using this decoder under several probabilistic assumptions. Furthermore, under the same assumptions, we estimate the probability of decreasing the error weight to a value that allows correction by the OSML decoder. This provides an upper bound on the decoding failure rate for the combination of a single iteration of parallel symbol flipping followed by the OSML decoder.

3.1 Distribution of Counters

Below we give necessary results on probabilistic distributions of syndrome values and counters $\sigma_{j,i}$, $j \in [\![1, n]\!]$, $i \in [\![0, q-1]\!]$, required for further analysis of decoding iteration of proposed parallel symbol–flipping decoder. Our analysis will rely on several assumptions that are analogous to those used in [5,45].

Assumption 1. *Let H be a parity–check of a random QC–MDPC code C in block–circulant form. It is assumed that each row of H is well modeled as a sample from uniform distribution over $\mathbb{F}_q^n$.*

Proposition 2. *Let $x \in \mathbb{F}_q^n$, $y \in \mathbb{F}_q^n$ be uniformly sampled. Let*

$$A_m = \Pr[\langle x, y\rangle \neq 0 \mid |\mathrm{supp}(x) \cap \mathrm{supp}(y)| = m].$$

Then A_m can be found recursively using

$$A_m = \begin{cases} (1 - A_{m-1}) + \frac{q-2}{q-1} A_{m-1}, & m \geq 1 \\ 0, & m = 0. \end{cases}$$

Proof. Without loss of generality, we assume that $\text{supp}(x) \cap \text{supp}(y) = \{1, \dots, m\}$. It follows that

$$\begin{aligned} A_m &= \Pr\left[\left(\sum_{i=1}^{m-1} x_i y_i = 0\right), x_m y_m \neq 0\right] + \Pr\left[\left(\sum_{i=1}^{m-1} x_i y_i \neq 0\right), x_m y_m \neq -\sum_{i=1}^{m-1} x_i y_i\right] = \\ &= \Pr\left[\left(\sum_{i=1}^{m-1} x_i y_i = 0\right)\right] \cdot \Pr\left[x_m y_m \neq 0 \mid \left(\sum_{i=1}^{m-1} x_i y_i = 0\right)\right] + \\ &+ \Pr\left[\left(\sum_{i=1}^{m-1} x_i y_i \neq 0\right)\right] \cdot \Pr\left[x_m y_m \neq -\alpha \mid \left(\sum_{i=1}^{m-1} x_i y_i = \alpha \neq 0\right)\right] = \\ &= (1 - A_{m-1}) \cdot 1 + A_{m-1} \frac{q-2}{q-1}. \end{aligned}$$

Theorem 1. *Let $H = (h_{i,j})$ be a parity-check matrix of (γ, δ)-regular QC-MDPC code C of length n. Let $e \in \mathbb{F}_q^n$ be a random error of weight t, and $s = eH^\mathsf{T}$ be its syndrome. Then for any row h_i of H, such that $j \in \text{supp}(h_i)$*

$$\Pr[s_i h_{i,j}^{-1} = e_j \mid e_j \neq 0] = \sum_{i=0}^{\min(\delta-1, t-1)} \frac{\binom{\delta-1}{i}\binom{n-\delta}{t-i-1}}{\binom{n-1}{t-1}} (1 - A_i), \tag{6}$$

$$\Pr[s_i h_{i,j}^{-1} = e_j \mid e_j = 0] = \sum_{i=0}^{\min(\delta-1, t)} \frac{\binom{\delta-1}{i}\binom{n-\delta}{t-i}}{\binom{n-1}{t}} (1 - A_i), \tag{7}$$

$$\Pr[s_i h_{i,j}^{-1} = \alpha \neq e_j \mid e_j \neq 0] = (q-1)^{-1} \left(1 - \Pr[s_i h_{i,j}^{-1} = e_j \mid e_j \neq 0]\right), \tag{8}$$

$$\Pr[s_i h_{i,j}^{-1} = \alpha \neq 0 \mid e_j = 0] = (q-1)^{-1} \left(1 - \Pr[s_i h_{i,j}^{-1} = e_j \mid e_j = 0]\right). \tag{9}$$

Proof. Since $j \in \text{supp}(h_i)$, Eq. (3) implies that $s_i h_{i,j}^{-1} = e_j + h_{i,j}^{-1} \langle e', h' \rangle$, where

$$e' = e_{[\![1,n]\!] \setminus \{j\}}, \quad h' = H_{i, [\![1,n]\!] \setminus \{j\}}.$$

One can easily note that

$$\text{wt}(e') = \begin{cases} t, & e_j = 0 \\ t-1, & e_j \neq 0 \end{cases}, \quad \text{wt}(h') = \delta - 1. \tag{10}$$

Since $s_i h_{i,j}^{-1} = e_j$ if and only if $\langle h', e' \rangle = 0$, it follows that

$$\Pr[s_i h_{i,j}^{-1} = e_j \mid e_j = \alpha] = \Pr[\langle e', h' \rangle = 0].$$

So, using Assumption 1, we obtain

$$\begin{aligned}\Pr[\langle e', h'\rangle = 0] &= \sum_{i=0}^{\min(\mathrm{wt}(e'),\mathrm{wt}(h'))} \Pr[\langle e', h'\rangle = 0,\ |\mathrm{supp}(e') \cap \mathrm{supp}(h')| = i] = \\ &= \sum_{i=0}^{\min(\mathrm{wt}(e'),\mathrm{wt}(h'))} (1 - A_i) \cdot \Pr[|\mathrm{supp}(e') \cap \mathrm{supp}(h')| = i] = \\ &= \sum_{i=0}^{\min(\mathrm{wt}(e'),\mathrm{wt}(h'))} \frac{\binom{\mathrm{wt}(h')}{i}\binom{n-1-\mathrm{wt}(h')}{\mathrm{wt}(e')-i}}{\binom{n-1}{\mathrm{wt}(e')}}(1 - A_i).\end{aligned}$$

Substituting (10) into this formula, we obtain (6) and (7). In addition, when $\langle e', h'\rangle \neq 0$, the product $\langle e', h'\rangle$ can assume any non–zero element of $\mathbb{F}_q$ with equal probabilities. Consequently, we obtain (8) and (9).

In the parallel symbol flipping decoder (see Algorithm 2), we propose the following flipping criterion based on counter values, using three decoding thresholds: th_O, th_E, and th_D. Namely, the position j of the received noisy codeword $z = c + e$ will be updated to $z_j - \alpha_i$ if the following conditions are satisfied:

1. $\sigma_{j,i} > \sigma_{j,\omega}$ for all $\omega \in [\![0, q-1]\!] \setminus \{i\}$, and thus $\sigma_j^* = \sigma_{j,i}$,
2. $\sigma_j^* \geq th_E$,
3. $\sigma_{j,0} < th_O$,
4. $\sigma_j^* - \sigma_{j,0} \geq th_D$.

Note that conditions 1–4 can be replaced by the single condition

$$\sigma_j = (\sigma_{j,0}, \ldots, \sigma_{j,q-1}) \in \Delta_{th_O,th_E,th_D}(i),$$

where $\Delta_{th_O,th_E,th_D}(i)$ is defined as follows

$$\begin{aligned}\Delta_{th_O,th_E,th_D}(i) = \Delta(i) = \Big\{(b_0, \ldots, b_{q-1}) \in \mathbb{Z}^q \mid \sum_{\omega=0}^{q-1} b_\omega = \gamma,\ b_i > \max_{\omega \neq i} b_z, \\ b_0 \leq th_O,\ b_i \geq th_E,\ b_i - b_0 \geq th_D\Big\}.\end{aligned}$$

In the following theorem, we will estimate the probability that the flipping criterion accurately determines the positions and values of errors.

Assumption 2. *We assume that the probability* $\Pr[\sigma_j \in \Delta(i)]$ *to flip position* j *to value* $z_j - \alpha_i$ *is a function only of error weight, i.e. it does not depend on error structure and the location* j.

Theorem 2. *Let* H *be a parity-check matrix of* (γ, δ)*–regular QC-MDPC code* C *of length* n *and dimension* k. *Let* $e \in \mathbb{F}_q^n$ *be a random error of weight* t. *Define*

$$p_1 = \Pr[s_i h_{i,j}^{-1} = e_j \mid e_j \neq 0], \quad p_2 = \Pr[s_i h_{i,j}^{-1} = \alpha \neq e_j \mid e_j \neq 0],$$

$$p_3 = \Pr[s_i h_{i,j}^{-1} = e_j \mid e_j = 0], \quad p_4 = \Pr[s_i h_{i,j}^{-1} = \alpha \neq e_j \mid e_j = 0].$$

Then the probability that non-zero error value will be estimated correctly is

$$p_{e\to c}(t) = \Pr[\sigma_j \in \Delta(i) \mid e_j = \alpha_i \neq 0] = \sum_{(b_0,\ldots,b_{q-1})\in\Delta(i)} \frac{\gamma!}{b_0!\ldots,b_{q-1}!} p_1^{b_i} p_2^{\gamma-b_i}, \tag{11}$$

and the probability of incorrect estimate in non-erroneous position is

$$p_{c\to e}(t) = (q-1)\cdot \Pr[\sigma_j \in \Delta(i) \mid e_j = 0], \tag{12}$$

where

$$\Pr[\sigma_j \in \Delta(i) \mid e_j = 0] = \sum_{(b_0,\ldots,b_{q-1})\in\Delta(i)} \frac{\gamma!}{b_0!\ldots,b_{q-1}!} p_3^{b_0} p_4^{\gamma-b_0}, \quad i \neq 0.$$

Proof. From Assumption 2 it follows that the probability

$$\Pr\left[\sigma_j = (b_0,\ldots,b_{q-1}) \mid e_j = \alpha_i \neq 0\right]$$

can be modelled using multinomial distribution with parameters

$$\left(\Pr[s_i h_{i,j}^{-1} = 0 \mid e_j \neq 0],\ldots,\Pr[s_i h_{i,j}^{-1} = \alpha_{q-1} \mid e_j \neq 0]\right) = (\underbrace{p_2,\ldots,p_2}_{i-1}, p_1, \underbrace{p_2,\ldots,p_2}_{q-i}).$$

Hence

$$\Pr\left[\sigma_j = (b_0,\ldots,b_{q-1}) \mid e_j = \alpha_i \neq 0\right] = \frac{\gamma!}{b_0!\ldots,b_{q-1}!} p_1^{b_i} p_2^{\gamma-b_i},$$

which implies (11). By similar reasoning, we can also obtain (12).

3.2 Analysis of Parallel Symbol-Flipping Decoder

In this subsection, we employ results of previous subsection to give an plausibility analysis of the one–step parallel symbol flipping decoder (Algorithm 2) and its combination with OSML decoder (Algorithm 3).

Algorithm 2: 1–iteration parallel symbol flipping decoder

```
Input: syndrome s = He^T
Output: estimated error ẽ
ẽ ← 0^n ∈ F_q^n;
for j ← 1 to n do
    using (4), (5) compute σ_j = (σ_{j,0}, ..., σ_{j,q-1}) and σ_j^*;
    if σ_j ∈ Δ(s) then
        ẽ_j ← ẽ_j + α_s
end
return ẽ
```

Note that, after each iteration some error positions can be estimated correctly and some non–erroneous positions can be estimated to be erroneous incorrectly. In the following proposition, we provide an analysis of the probability that 1-iteration version of this decoder transforms a random error e of weight t into some new error e' of weight t'.

Proposition 3. *Let e be a random error of weight t, then after execution Algorithm 2*

1. *the probability to correctly estimate u error positions from e is*

$$P_{correct}(t,u) = \binom{t}{u} (p_{e\to c}(t))^u (1 - p_{e\to c}(t))^{t-u},$$

2. *the probability to corrupt v non–erroneous positions is*

$$P_{corrupt}(t,v) = \binom{n-t}{v} (p_{c\to e}(t))^v (1 - p_{c\to e})^{n-t-v},$$

3. *the probability to transform e into an error e′ of weight t′ is*

$$\Pr(t \to t') = \sum_{t-u+v=t'} P_{correct}(t,u) P_{corrupt}(t,v).$$

Proof. Assumption 2 implies that the flip decisions are statistically independent and depend solely on the error weight. It follows that $P_{correct}(t,u)$ and $P_{corrupt}(t,v)$ can be modeled as samples from binomial distributions with parameters $p_{e\to c}(t)$ and $p_{c\to e}(t)$ described in Theorem 2, respectively. The last claim trivially follows from the first two.

Corollary 3. *The decoding failure rate of* 1*-iteration parallel symbol-flipping decoder can be estimated as follows*

$$DFR_1 = 1 - \Pr(t \to 0).$$

Note that the new error e' is not random anymore and, therefore, the same analysis for further iteration is not possible. However, it is possible to decode e' using OSML decoder, which rely on no probabilistic assumptions.

Algorithm 3: PSF+OSML

```
Input: syndrome s = He^T
Output: estimated error ẽ
ẽ ← 0^n ∈ F_q^n;
for j ← 1 to n do
    using (4), (5) compute σ_j = (σ_{j,0}, ..., σ_{j,q-1}) and σ_j^*;
    if σ_j ∈ Δ(s) then
        ẽ_j ← ẽ_j + α_s
end
s ← He^T − Hẽ^T;
ẽ ← ẽ + OSML(s);
return ẽ
```

Thus, we obtain the following corollary:

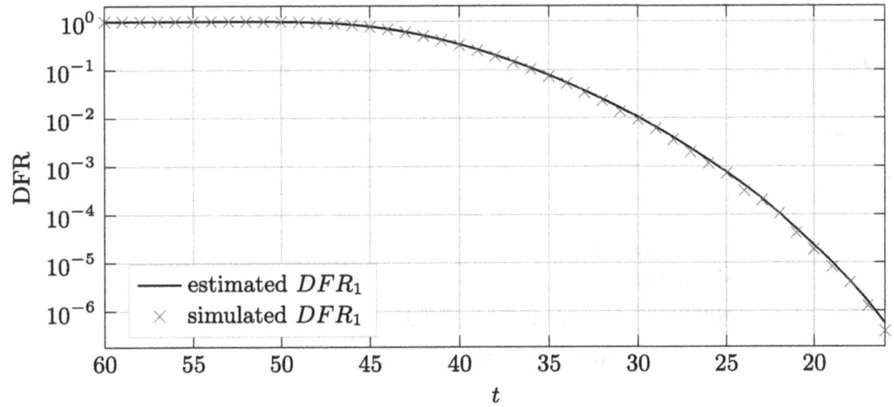

Fig. 2. Simulation results of DFR_1 for random QC-MDPC $[n = 2 \cdot 2339, k = 2339]_4$–codes over $\mathbb{F}_4$ ($l = 2$, $p = 2339$, $\gamma = 37$), with decoding thresholds $(th_O, th_E, th_D) = (18, 4, 4)$

Corollary 4. *Let e be a random error of weight t, let τ be the number of errors which can be corrected with certainty using OSML decoder. Then DFR of this combination is upper bounded by*

$$DFR_2 = 1 - \sum_{t'=0}^{\tau} \Pr(t \to t').$$

In Figs. 2, 3, 4, we present the results of numerical simulations and compare them with the obtained theoretical estimates. Each experiment involved generating a random key and decoding a random error. For each error weight, the experiments were conducted until 100 decoding failures were detected or until 10^8 experiments were performed, whichever occurred first.

We observe that the theoretical estimates of DFR_1 and DFR_2 closely match the simulation results, substantiating the accuracy of the obtained theoretical model.

4 Choice of Cryptosystem Parameters

The choice of parameters of QC–MDPC cryptosystems is determined by the complexity of potential attacks on such cryptosystems. Specifically, the parameters of the cryptosystem should be chosen in such a way that the best key-recovery attacks and message-recovery attacks require a sufficiently large number of operations.

The most effective message–recovery attacks are a family of information set decoding (ISD) algorithms, designed for decoding random codes. This family includes the Prange algorithm, the Lee-Brickell algorithm, Stern algorithm, BJMM, ball–collision, etc. An overview of ISD–algorithms can be found in [49].

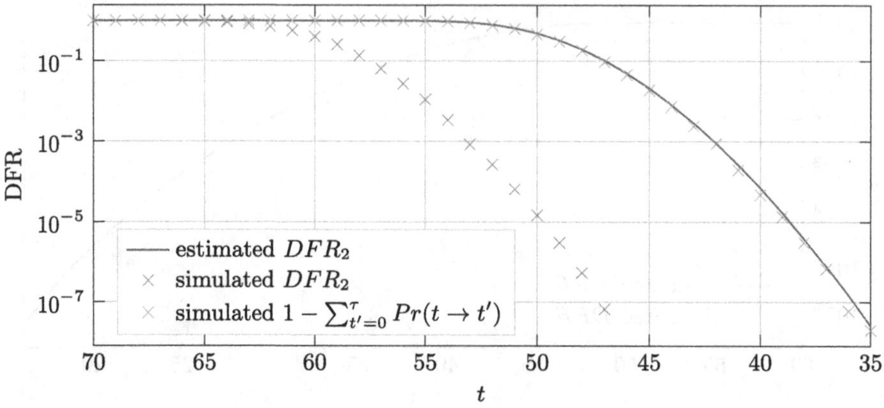

Fig. 3. Simulation results of DFR_2 for random QC-MDPC $[n = 2 \cdot 2339, k = 2339]_4$–codes over $\mathbb{F}_4$ ($l = 2$, $p = 2339$, $\gamma = 37$, $(th_O, th_E, th_D) = (18, 4, 4)$, and $\tau = 4$). For each experiment, we generated a random code and then checked if its OSML bound (see Corollary 2) is $\geq \tau$. If a code had a lower bound, it was rejected. We chose $\tau = 4$ to reject no more than 50% of keys (the actual rejection rate was 3%).

The average complexity of these algorithms can be directly estimated using a formula that depends on parameters such as the field size q, code length n, code dimension k, and the weight of the error w that needs to be found. For non-binary code direct complexity estimates for the Lee-Brickell and Stern algorithms can be found in [49], for BJMM in [25], and for ball-collision in [27].

It should be noted that for quasi-cyclic codes of order n', it has been shown [41] that the complexity of ISD attacks can be reduced by a factor of $\sqrt{n'}$ compared to codes without any structure. One of the features of QC-MDPC cryptosystems is that for key-recovery attacks, which involve finding low-weight dual codewords, the best attacks are also based on ISD. This is because the same algorithms can easily be adapted to search for codewords of a given weight instead of finding an error of a given weight. For quasi-cyclic codes, in this case, it is also possible to reduce the complexity by a factor of n' compared to random codes.

Furthermore, we must consider the decoding failure rate since in [26], Q. Guo et al. proposed a *reaction attack* that allows the recovery of secret keys in cryptosystems based on *binary QC-MDPC codes* by exploiting decoding failures. The original description assumes that $l = 2$, i.e., $n = 2n'$, but it can be easily generalized to other cases. This attack is based on the observation that certain error patterns are more easily decodable than other ones. Namely, let $\mathcal{E}_r$ be the set of error patterns of the following form:

$$\mathcal{E}_r = \{(e, \mathbf{0}) \in \mathbb{F}_2^{2p} \mid e \in \mathbb{F}_2^p,\ \exists \text{ distinct } s_1, s_2, \ldots, s_t, \text{ s.t. } e_{s_i} = 1 \text{ and } s_{2i} = (s_{2i-1} + r) \bmod n' \text{ for } i \in [\![1, t/2]\!]\}$$

Let $\mathbf{h_1} \in \mathbb{F}_q^{n'}$ denote the first row of H_1 (see (2)). Let $\psi(r)$ denote the number of pairs of non-zero positions of $\mathbf{h}_1$ placed at distance d. The distance between

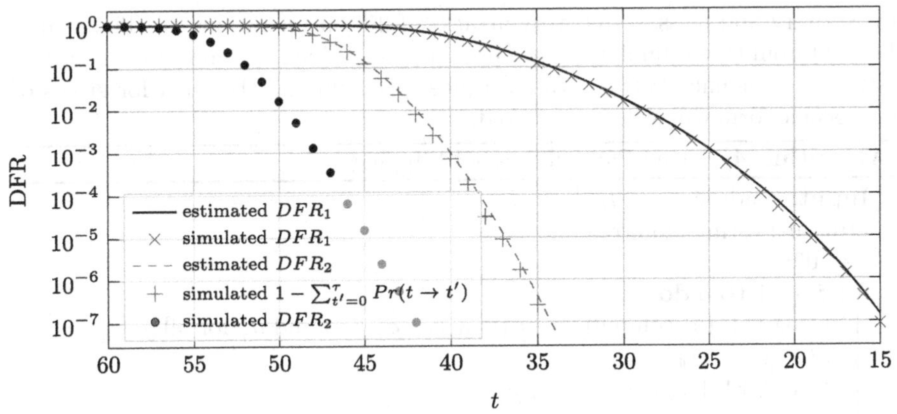

Fig. 4. Simulation results of DFR_1 and DFR_2 for random QC–MDPC $[n = 2 \cdot 1583, k = 1583]_8$–codes over $\mathbb{F}_8$ ($l = 2$, $p = 1583$, $\gamma = 37$, $(th_0, th_E, th_D) = (18, 4, 4)$, and $\tau = 4$)

i and j is computed as $\min\{(i-j) \bmod n', (j-i) \bmod n'\}$. The set of values $\psi(i)$, $i \in [\![1, \lfloor n'/2 \rfloor]\!]$, is called the *distance spectrum* of $\mathbf{h_1} \in \mathcal{R}_n$. In [26], it was shown that there is a correlation between the decoding failure rate on errors from $\mathcal{E}_r$ and the value of $\psi(r)$. Specifically, the larger $\psi(r)$ is, the lower the DFR for errors from $\mathcal{E}_r$.

Therefore, computing the DFR on errors from $\mathcal{E}_r$ for different r allows for the recovery of the distance spectrum of $\mathbf{h_1}$ and subsequently $\mathbf{h_1}$ itself. Consequently, it becomes possible to reconstruct the secret key of binary QC-MDPC cryptosystems by exploiting the decoding failures. Below, we demonstrate how this attack can be applied to the non-binary case as well.

Table 1. Dependency between simulated DFR for random errors $e \in \tilde{\mathcal{E}_r}$ and the values $\psi(r)$. The results are averaged over 100 random QC–MDPC [4678, 2339]–codes.

$\psi(r)$	0	1	2	3	4
simulated DFR	0.0203	0.0134	0.0085	0.0059	0.0039

In our experiments, we observed a correlation between the DFR for errors from $\tilde{\mathcal{E}r}$ and the values of $\psi(r)$, where the set $\tilde{\mathcal{E}_r}$ is defined as follows:

$$\tilde{\mathcal{E}_r} = \{(e, \mathbf{0}) \in \mathbb{F}_q^{2p} \mid e \in \mathbb{F}_2^p,\ \exists \text{ distinct } s_1, s_2, \ldots, s_t, \text{ s.t. } e_{s_i} \neq 0 \text{ and } s_{2i} = (s_{2i-1} + r) \bmod n' \text{ for } i \in [\![1, t/2]\!]\}$$

For instance, we conducted simulations to decode errors of weight $t = 84$ from $\tilde{\mathcal{E}_r}$ using Algorithm 4 for random QC-MDPC codes over $\mathbb{F}_4$ with parameters $n' = 2339$, $l = 2$, and $\gamma = 37$, which ensure a minimal cost of ISD-based key-recovery and message-recovery attacks of 2^{80} bit operations [9]. The results obtained from these simulations are presented in Table 1. As shown in the table, a strong dependency between the distance spectrum and the DFR for errors of this specific form can still be observed.

Algorithm 4: Sorted Parallel Symbol Flipping

```
Input: syndrome s = He^T
Output: estimated error ẽ
ẽ ← 0^n;
for it ← 1 to 5 do
    using (4), (5) compute σ_j = (σ_{j,0}, ..., σ_{j,q-1}) and σ*_j for all j ∈ ⟦1, n⟧,
      i ∈ ⟦0, q − 1⟧ ;
    th ← 20th_largest(σ*_j − σ_{j,0});
    for j ← 1 to n do
        if σ_j − σ_{j,0} ≥ max(th, 1) then
            i* ← argmax_{i∈⟦1,q−1⟧} σ_{i,j};
            ẽ_j ← ẽ_j + α_{i*};
    end
    s ← He^T − Hẽ^T;
    if s = 0 then return ẽ;
end
return fail;
```

Thus, it is possible to reconstruct the support of the secret vector $\mathbf{h_1}$ (up to a cyclic shift) using the following steps:

1. for each $r \in [\![1, \lfloor n'/2 \rfloor]\!]$ numerically estimate DFR for random errors from $\tilde{\mathcal{E}}_r$, and then use the obtained results to recover the distance spectrum ψ of $\mathbf{h_1}$;
2. recover $\text{supp}(h_1)$ using the procedure described in [26] for finding positions of ones in $\mathbf{h_1}$ for the binary case

Once $\text{supp}(\mathbf{h_1})$ is recovered, it is possible to recover the whole secret key $(\mathbf{h_1}, \mathbf{h_2})$ in the non-binary case as follows. Let I be an information set such that

$$|I \cap \text{supp}(\mathbf{h_1} \mid \mathbf{h_2})| = 1,$$

then the matrix $\tilde{H}_{:,I}^{-1}\tilde{H} = H_{:,I}^{-1}H$ contains the row $(\mathbf{h_1}, \mathbf{h_2})$ or its quasi–circular shift. When $\text{supp}(\mathbf{h_1})$ is known, I can be constructed of one element from $\text{supp}(\mathbf{h_1})$, $n' - \gamma$ elements from $[\![1, n']\!] \setminus \text{supp}(\mathbf{h_1})$, and randomly guessed $\gamma - 1$ elements from $[\![n'+1, 2n']\!]$. Therefore, the probability of finding a suitable I can be estimated as follows:

$$\binom{n' - \gamma}{\gamma - 1} \cdot \binom{n'}{\gamma - 1}^{-1}.$$

So, the method described above in our experiments allowed reconstruction of secret key with significantly lower complexity than claimed security level of 2^{80} bit operations.

It follows that, when choosing the parameters of QC–MDPC cryptosystem that can be converted into IND–CCA2 secure KEM in non–binary case the design criteria are the complexity of ISD–based key–recovery, and message–recovery attacks and small enough decoding failure rate making reaction attacks infeasible. Table 2 provides potential parameters of QC-MDPC cryptosystems over $\mathbb{F}_4$, with $l = 2$ and $n' = p$ being a prime such that the polynomial $x^p - 1$ has a low number of irreducible factors. These parameters are given for three different security levels: $\lambda \in \{128, 192, 256\}$, which correspond to the complexity of breaking AES with the corresponding key sizes. All the proposed instances are designed to have $DFR_2 \leq 2^{-\lambda}$ (see Corollary 4). Note that the resulting public key sizes (pk_{size}) are slightly larger than in the binary case (28, 277, 52, 667, 83, 579 respectively [5]). Moreover, increasing the field size to $q = 8$ with security level $\lambda = 128$ yields an estimated public key size of 36, 321 bits ($p = 12,107$, $\gamma = 69$, $t = 130$). Thus, for a fixed security level, public key size grows with increasing field size. Indeed, to maintain the same or smaller pk_{size} when increasing q, one must consider shorter MDPC codes. However, due to the complexity of ISD-based key–recovery and message–recovery attacks, γ and t are nearly the same across various ranges of q, implying higher-density codes. Therefore, the increased field size does not appear to compensate for the negative impact of increased code density.

Table 2. Cryptosystem parameters

q	λ	p	γ	t	(th_0, th_E, th_D)	τ	pk_{size} (bits)
4	128 (2^{143} *bit operations*)	16 651	71	132	$(\gamma, 5, 5)$	9	33, 302
4	192 (2^{207} *bit operations*)	30 971	103	197	$(\gamma, 6, 6)$	12	61, 942
4	256 (2^{272} *bit operations*)	47 903	137	263	$(\gamma, 6, 6)$	16	95, 806

5 Conclusion

In this paper, we have studied the guaranteed error-correction capability of the one-step majority logic (OSML) decoder and provided a plausibility analysis of the 1-iteration parallel symbol flipping decoder for non-binary QC-MDPC codes. Through this analysis, we were able to estimate the decoding failure rate (DFR) of the combined use of these decoders, where parallel symbol flipping is employed to reduce the error weight to a level at which the OSML decoder can successfully correct any remaining errors. Consequently, we have obtained worst-case estimates of the DFR, considering some minimalistic and reasonable assumptions. The accuracy and validity of our theoretical model have been verified through numerical simulations.

Furthermore, we have demonstrated the importance of considering key-recovery reaction attacks when designing non-binary QC-MDPC cryptosystems.

This implies that such cryptosystems need to be constructed with an extremely low DFR in order to achieve IND-CCA2 security with long-term keys. Finally, we have provided possible parameters for different NIST security levels of non-binary QC-MDPC cryptosystems, along with their theoretically estimated DFR.

It should be noted that the resulting key sizes are slightly larger than those in the binary case. Therefore, it appears that using non-binary QC-MDPC codes does not offer any benefits in terms of reducing the public-key sizes of IND-CCA2-secure cryptosystems considering the reaction attack. However, there is a possibility that replacing the quasi-cyclic structure with a more general (non-abelian) quasi-group structure, specifically replacing circulant matrices with matrices of multiplication operators in group algebras, could potentially hinder the reaction attack.

Additionally, by abandoning the requirement of key re-usage, it becomes possible to consider more sophisticated decoders for cryptosystems resistant against chosen plaintext attacks (CPA–secure). The study of such decoders can only be carried out through experimental methods and may provide benefits in terms of reducing key sizes, as previously explored in [9].

It is worth mentioning that the obtained in this paper theoretical models could potentially be useful for providing conservative estimates of the DFR of non-binary codes in telecommunications applications.

References

1. Alagic, G., et al.: Status report on the third round of the NIST post-quantum cryptography standardization process. US Department of Commerce, NIST (2022). https://doi.org/10.6028/NIST.IR.8413
2. Apon, D., Perlner, R., Robinson, A., Santini, P.: Cryptanalysis of LEDAcrypt. In: Micciancio, D., Ristenpart, T. (eds.) CRYPTO 2020, Part III. LNCS, vol. 12172, pp. 389–418. Springer, Cham (2020). https://doi.org/10.1007/978-3-030-56877-1_14
3. Aragon, N., et al.: Bike: bit flipping key encapsulation. bikesuite.org
4. Arpin, S., Billingsley, T.R., Hast, D.R., Lau, J.B., Perlner, R., Robinson, A.: A study of error floor behavior in QC-MDPC codes. In: Cheon, J.H., Johansson, T. (eds.) Post-Quantum Cryptography. Lecture Notes in Computer Science, vol. 13512, pp. 89–103. Springer, Cham (2022). https://doi.org/10.1007/978-3-031-17234-2_5
5. Baldi, M., Barenghi, A., Chiaraluce, F., Pelosi, G., Santini, P.: Analysis of in-place randomized bit-flipping decoders for the design of LDPC and MDPC code-based cryptosystems. In: Obaidat, M.S., Ben-Othman, J. (eds.) ICETE 2020. CCIS, vol. 1484, pp. 151–174. Springer, Cham (2021). https://doi.org/10.1007/978-3-030-90428-9_7
6. Baldi, M., Barenghi, A., Chiaraluce, F., Pelosi, G., Santini, P.: Performance bounds for QC-MDPC codes decoders. In: Wachter-Zeh, A., Bartz, H., Liva, G. (eds.) Code-Based Cryptography. Lecture Notes in Computer Science, vol. 13150, pp. 95–122. Springer International Publishing, Cham (2022). https://doi.org/10.1007/978-3-030-98365-9_6

22. Gabidulin, E.M., Paramonov, A.V., Tretjakov, O.V.: Ideals over a non-commutative ring and their application in cryptology. In: Davies, D.W. (ed.) EUROCRYPT 1991. LNCS, vol. 547, pp. 482–489. Springer, Heidelberg (1991). https://doi.org/10.1007/3-540-46416-6_41
23. Gaborit, P.: Shorter keys for code based cryptography. In: Proceedings of the 2005 International Workshop on Coding and Cryptography (WCC 2005), pp. 81–91 (2005)
24. Gallager, R.: Low-density parity-check codes. IRE Trans. Inf. Theory **8**(1), 21–28 (1962)
25. Gueye, C.T., Klamti, J.B., Hirose, S.: Generalization of BJMM-ISD using may-Ozerov nearest neighbor algorithm over an arbitrary finite field $\mathbb{F}_q$. In: El Hajji, S., Nitaj, A., Souidi, E.M. (eds.) C2SI 2017. LNCS, vol. 10194, pp. 96–109. Springer, Cham (2017). https://doi.org/10.1007/978-3-319-55589-8_7
26. Guo, Q., Johansson, T., Wagner, P.S.: A key recovery reaction attack on QC-MDPC. IEEE Trans. Inf. Theory **65**, 1845–1861 (2019). https://doi.org/10.1109/TIT.2018.2877458
27. Interlando, C., Khathuria, K., Rohrer, N., Rosenthal, J., Weger, V.: Generalization of the ball-collision algorithm. arXiv preprint: arXiv:1812.10955 (2018)
28. Ivanov, F., Krouk, E., Zyablov, V.: New code-based cryptosystem based on binary image of generalized reed-Solomon code. In: 2021 XVII International Symposium "Problems of Redundancy in Information and Control Systems"(REDUNDANCY), pp. 66–69. IEEE (2021)
29. Janwa, H., Moreno, O.: McEliece public key cryptosystems using algebraic-geometric codes. Des. Codes Crypt. **8**(3), 293–307 (1996)
30. Kosolapov, Y., Lelyuk, A.: Cryptanalysis of the BBCRS system on reed-muller binary codes. Bull. South Ural State Univ. Ser. Math. Modell. Program. Comput. Softw. **14**, 18–32 (2021). https://doi.org/10.14529/mmp210302
31. McEliece, R.J.: Public-key cryptosystem based on algebraic coding theory. PL Deep Space Netw. Prog. Report **42**, 114–116 (1978)
32. Minder, L., Shokrollahi, A.: Cryptanalysis of the Sidelnikov cryptosystem. In: Naor, M. (ed.) EUROCRYPT 2007. LNCS, vol. 4515, pp. 347–360. Springer, Heidelberg (2007). https://doi.org/10.1007/978-3-540-72540-4_20
33. Misoczki, R., Tillich, J.P., Sendrier, N., Barreto, P.S.L.M.: MDPC-McEliece: new McEliece variants from moderate density parity-check codes, pp. 2069–2073. IEEE (2013). https://doi.org/10.1109/ISIT.2013.6620590
34. Monico, C., Rosenthal, J., Shokrollahi, A.: Using low density parity check codes in the McEliece cryptosystem, pp. 215. IEEE (2000). https://doi.org/10.1109/ISIT.2000.866513
35. Niederreiter, H.: Knapsack-type cryptosystems and algebraic coding theory. Prob. Control Inf. Theory **15**, 159–166 (1986)
36. Otmani, A., Tillich, J.P., Dallot, L.: Cryptanalysis of two McEliece cryptosystems based on quasi-cyclic codes. Math. Comput. Sci. **3**, 129–140 (2010). https://doi.org/10.1007/s11786-009-0015-8
37. Overbeck, R.: Statistical decoding revisited. In: Batten, L.M., Safavi-Naini, R. (eds.) ACISP 2006. LNCS, vol. 4058, pp. 283–294. Springer, Heidelberg (2006). https://doi.org/10.1007/11780656_24
38. Overbeck, R.: Structural attacks for public key cryptosystems based on Gabidulin codes. J. Cryptol. **21**(2), 280–301 (2008)
39. Santini, P., Battaglioni, M., Baldi, M., Chiaraluce, F.: Analysis of the error correction capability of LDPC and MDPC codes under parallel bit-flipping decoding

7. Baldi, M., Bianchi, M., Chiaraluce, F., Rosenthal, J., Schipani, D.: Enhanced publ key security for the McEliece cryptosystem. J. Cryptol. **29**(1), 1–27 (2014). https:/ doi.org/10.1007/s00145-014-9187-8
8. Baldi, M., Bodrato, M., Chiaraluce, F.: A new analysis of the McEliece cryptosys tem based on QC-LDPC codes. In: Ostrovsky, R., De Prisco, R., Visconti, I. (eds.) SCN 2008. LNCS, vol. 5229, pp. 246–262. Springer, Heidelberg (2008). https:// doi.org/10.1007/978-3-540-85855-3_17
9. Baldi, M., Cancellieri, G., Chiaraluce, F., Persichetti, E., Santini, P.: Using non-binary LDPC and MDPC codes in the McEliece cryptosystem. In: 2019 AEIT International Annual Conference (AEIT), pp. 1–6. IEEE (2019)
10. Baldi, M., Chiaraluce, F.: Cryptanalysis of a new instance of McEliece cryptosystem based on QC-LDPC codes. In: 2007 IEEE International Symposium on Information Theory, pp. 2591–2595. IEEE (2007)
11. Baldi, M., Chiaraluce, F., Garello, R., Mininni, F.: Quasi-cyclic low-density parity-check codes in the McEliece cryptosystem. In: 2007 IEEE International Conference on Communications, pp. 951–956. IEEE (2007)
12. Berger, T.P., Cayrel, P.-L., Gaborit, P., Otmani, A.: Reducing key length of the McEliece cryptosystem. In: Preneel, B. (ed.) AFRICACRYPT 2009. LNCS, vol. 5580, pp. 77–97. Springer, Heidelberg (2009). https://doi.org/10.1007/978-3-642-02384-2_6
13. Berlekamp, E., McEliece, R., van Tilborg, H.: On the inherent intractability of certain coding problems (corresp.). IEEE Trans. Inf. Theory **24**, 384–386 (1978). https://doi.org/10.1109/TIT.1978.1055873
14. Bernstein, D.J., Lange, T.: Post-quantum cryptography. Nature **549**, 188–194 (2017). https://doi.org/10.1038/nature23461
15. Borodin, M.A., Chizhov, I.V.: Effective attack on the McEliece cryptosystem based on reed-muller codes. Discret. Math. Appl. **24**(5), 273–280 (2014)
16. Couvreur, A., Lequesne, M., Tillich, J.-P.: Recovering short secret keys of RLCE in polynomial time. In: Ding, J., Steinwandt, R. (eds.) PQCrypto 2019. LNCS, vol. 11505, pp. 133–152. Springer, Cham (2019). https://doi.org/10.1007/978-3-030-25510-7_8
17. Couvreur, A., Márquez-Corbella, I., Pellikaan, R.: Cryptanalysis of public-key cryptosystems that use subcodes of algebraic geometry codes. In: Pinto, R., Malonek, P.R., Vettori, P. (eds.) Coding Theory and Applications. CSMS, vol. 3, pp. 133–140. Springer, Cham (2015). https://doi.org/10.1007/978-3-319-17296-5_13
18. Couvreur, A., Otmani, A., Tillich, J.-P., Gauthier-Umaña, V.: A polynomial-time attack on the BBCRS scheme. In: Katz, J. (ed.) PKC 2015. LNCS, vol. 9020, pp. 175–193. Springer, Heidelberg (2015). https://doi.org/10.1007/978-3-662-46447-2_8
19. Deundyak, V.M., Kosolapov, Y.V., Maystrenko, I.A.: On the decipherment of Sidel'nikov-type cryptosystems. In: Baldi, M., Persichetti, E., Santini, P. (eds.) CBCrypto 2020. LNCS, vol. 12087, pp. 20–40. Springer, Cham (2020). https:// doi.org/10.1007/978-3-030-54074-6_2
20. Faugère, J.-C., Otmani, A., Perret, L., Tillich, J.-P.: Algebraic cryptanalysis of McEliece variants with compact keys. In: Gilbert, H. (ed.) EUROCRYPT 2010. LNCS, vol. 6110, pp. 279–298. Springer, Heidelberg (2010). https://doi.org/10.1007/978-3-642-13190-5_14
21. Fossorier, M.P., Lin, S.: Soft-decision decoding of linear block codes based on ordered statistics. IEEE Trans. Inf. Theory **41**(5), 1379–1396 (1995)

and application to cryptography. IEEE Trans. Commun. **68**, 4648–4660 (2020). https://doi.org/10.1109/TCOMM.2020.2987898
40. Sendrier, N.: On the structure of randomly permuted concatenated code. Ph.D. thesis, INRIA (1995)
41. Sendrier, N.: Decoding one out of many. In: Yang, B.-Y. (ed.) PQCrypto 2011. LNCS, vol. 7071, pp. 51–67. Springer, Heidelberg (2011). https://doi.org/10.1007/978-3-642-25405-5_4
42. Shor, P.W.: Algorithms for quantum computation: discrete logarithms and factoring. In: Proceedings 35th Annual Symposium on Foundations of Computer Science, pp. 124–134. IEEE (1994)
43. Sidelnikov, V.M.: A public-key cryptosystem based on binary reed-muller codes. Discret. Math. Appl. **4**(3), 191–208 (1994)
44. Sidelnikov, V.M., Shestakov, S.O.: On insecurity of cryptosystems based on generalized reed-solomon codes. Discrete Math. Appl. **2** (1992)
45. Tillich, J.P.: The decoding failure probability of MDPC codes, pp. 941–945. IEEE (2018). https://doi.org/10.1109/ISIT.2018.8437843
46. Vasseur, V.: Post-quantum cryptography: a study of the decoding of QC-MDPC codes. Ph.D. thesis, Université de Paris (2021)
47. Vedenev, K., Kosolapov, Y.: Cryptanalysis of Ivanov-Krouk-Zyablov cryptosystem. In: Deneuville, J.C. (ed.) Code-Based Cryptography. Lecture Notes in Computer Science, vol. 13839, pp. 137–153. Springer Nature Switzerland, Cham (2023). https://doi.org/10.1007/978-3-031-29689-5_8
48. Wang, Y.: Quantum resistant random linear code based public key encryption scheme RLCE. In: 2016 IEEE International Symposium on Information Theory (ISIT), pp. 2519–2523. IEEE (2016)
49. Weger, V.: Information set decoding in the lee metric and the local to global principle for densities. Ph.D. thesis, PhD thesis, University of Zurich (2020)
50. Wieschebrink, C.: Cryptanalysis of the Niederreiter public key scheme based on GRS subcodes. In: Sendrier, N. (ed.) PQCrypto 2010. LNCS, vol. 6061, pp. 61–72. Springer, Heidelberg (2010). https://doi.org/10.1007/978-3-642-12929-2_5

FuLeeca: A Lee-Based Signature Scheme

Stefan Ritterhoff, Georg Maringer(✉), Sebastian Bitzer, Violetta Weger, Patrick Karl, Thomas Schamberger, Jonas Schupp, and Antonia Wachter-Zeh

TUM School of Computation, Information and Technology, Technical University of Munich, Munich, Germany
{stefan.ritterhoff,georg.maringer,sebastian.bitzer,violetta.weger, patrick.karl,t.schamberger,jonas.schupp,antonia.wachter-zeh}@tum.de

Abstract. In this work, we introduce a new code-based signature scheme, called FuLeeca, based on the NP-hard problem of finding codewords of given Lee-weight. The scheme follows the Hash-and-Sign approach applied to quasi-cyclic codes. Similar approaches in the Hamming metric have suffered statistical attacks, which revealed the small support of the secret basis. Using the Lee metric, we are able to thwart such attacks. We use existing hardness results on the underlying problem and study adapted statistical attacks. We propose parameters for FuLeeca and compare them to an extensive list of proposed post-quantum secure signature schemes including the ones already standardized by NIST. This comparison reveals that FuLeeca is competitive. For example, for NIST category I, i.e., 160 bit of classical security, we obtain an average signature size of 1100 bytes and public key sizes of 1318 bytes. Comparing the total communication cost, i.e., the sum of the signature and public key size, we see that FuLeeca is only outperformed by Falcon while the other standardized schemes Dilithium and SPHINCS+ show higher communication costs than FuLeeca.

Keywords: Post-Quantum cryptography · Signature scheme · Code-Based cryptography · Lee metric

1 Introduction

Due to the threat coming from capable quantum computers, NIST initialized in 2016 a standardization call for post-quantum alternatives.

Since the standardization call several of the submitted cryptosystems have been broken or removed from the competition as the proposed parameters seemed inferior to other schemes. Recently, several cryptosystems have been selected for standardization, both for key encapsulation and digital signatures. The most competitive and already selected signature schemes in terms of parameter sizes (keys and signature sizes) are based on structured lattices, namely CRYSTALS-Dilithium [28] and Falcon [36]. If signature sizes or signing times are not a major concern, hash-based signatures like SPHINCS+ [5] provide even

A. Esser and P. Santini (Eds.): CBCrypto 2023, LNCS 14311, pp. 56–83, 2023.
https://doi.org/10.1007/978-3-031-46495-9_4

smaller key sizes. Since it is desirable to have a broader variety of schemes to choose from, NIST reopened the standardization call for digital signatures.

One possibility to build quantum-secure signature schemes is to rely on hard problems from coding theory which have been examined over decades [9,10,15,18]. While classical code-based cryptography considers vector spaces endowed with the Hamming metric, other metrics, such as the rank metric, have attracted attention in the context of cryptography and show great potential for smaller key sizes. To the best of our knowledge, this work marks the first Lee-based cryptographic primitive (note that the paper [42] introduces a McEliece framework in the Lee metric, but without a particular code to instantiate it and thus we do not consider it a cryptographic primitive).

In general, there are two main methods to construct code-based signature schemes: the first one applies the Fiat-Shamir transform [35] to a code-based zero-knowledge protocol and the second one is called Hash-and-Sign approach [14]. The former approach usually suffers from large signature sizes due to large cheating probabilities within the zero-knowledge protocol. At the same time, the latter features small signature sizes at the cost of larger public key sizes.

The signature scheme we present in this paper is based on the Hash-and-Sign approach, which was introduced in 2001 by Courtois, Finiasz and Sendrier [24] (following the idea of [14]). This so called CFS scheme is a direct adaption of the McEliece public-key encryption scheme. In fact, the rationale is to start with an algebraically structured secret code that comes with an efficient decoding algorithm. The public key is a disguised version of the secret code. To generate a signature, the message and a salt is hashed until the digest results in a syndrome of a low-weight error vector. This approach has some potential drawbacks that have been exploited for attacks in the past: on the one hand, the public code might be distinguishable from a random code and thus leak information on the secret code[1]. On the other hand, the event that the hash of a message is a syndrome of a low weight error vector is highly unlikely and therefore this process has to be repeated many times. This causes the signing time of CFS to be impractically high. Additionally, as the public key is a disguised version of an algebraically structured code, the public key size of CFS tends to be rather large. The CFS scheme was the starting point for several Hash-and-Sign signature schemes, such as [6,23,40], which have not survived cryptanalysis [46,48]. The code-based scheme WAVE [26] also follows the same blueprint but translated into the theoretical framework of [39]. Additionally, it is based on the hardness of finding errors having large Hamming weights instead of small ones, thereby preventing all aforementioned attacks and so far no successful cryptanalysis has been mounted.

Code-based signature schemes based on quasi-cyclic structures with low density codes in the Hamming metric, e.g., [6,47], are vulnerable to statistical key recovery attacks [27,51]. These attacks have in common that they make use of the small support of the secret key. An attacker can recover the sparse secret key

[1] See for example [31], where the CFS scheme using high rate Goppa codes has been attacked.

by observing the distribution of many signatures and comparing it to a random distribution. The use of the Lee metric thwarts such attacks, as even if the Lee weight of the secret basis is low, the number of non-zero entries is relatively large.

FuLeeca is based on quasi-cyclic codes in the Lee metric. In a nutshell, the signature scheme works as follows: the secret key is a quasi-cyclic generator matrix, where the generators have Lee weight according to the Lee-metric Gilbert-Varshamov (GV) bound and the public key is its systematic form. Note that recovering the original generators is as hard as the problem of finding codewords of given Lee weight, which has been proven to be NP-hard [56]. The binary hash output of the message $\boldsymbol{m}$ is mapped onto $\{\pm 1\}$ and is considered as the target vector $\boldsymbol{c}$ for the main step of the scheme: the signer uses the secret generators to find a codeword, which is connected to the signature for $\boldsymbol{m}$, with two properties: firstly, the Lee weight should be low and secondly, the signum of the codeword should have many 1s, respectively -1s, in the same places as the target vector $\boldsymbol{c}$. This second property, called sign matching, is used to bind the message to the signature, while the first property is essential for the scheme's security.

For our chosen parameters targeting NIST security level I, the public key and signature sizes are only 1318 bytes and 1100 bytes, respectively. The total size, which is the sum of the public key and signature sizes is often used for comparisons, since in certificates one would need to download both. The total size of FuLeeca is smaller than that of Dilithium [28] and SPHINCS$^+$ [5], and slightly larger than Falcon [36], which are the three signature schemes currently selected for standardization by NIST.

A multiple-use signature scheme should have an existential unforgeability under adaptive chosen message attacks (EUF-CMA) security proof. For code-based signatures constructed from a zero-knowledge protocol this property is assured by using a sufficiently large number of rounds with respect to the desired security level and the cheating probability of the zero-knowledge protocol. However, the EUF-CMA security proof is notoriously difficult for Hash-and-Sign approaches. To the best of our knowledge, WAVE [26] is the only known code-based Hash-and-Sign signature scheme that provides such a proof. Unfortunately, the achieved public-key size of more than 2 megabytes for 128 bit classical security is very large compared to Falcon's 897 bytes.

The capability of breaking FuLeeca (e.g. recovering the secret key from a polynomial number of collected signatures) is crucially based on the fact that signatures do not leak any useful information on the secret key. An EUF-CMA security proof would immediately follow by assuming that such problem is hard. However, we feel like such a proof would be immature, given the current state of affairs. Indeed, such problem is somewhat non-standard, since this problem has not been used in cryptography before. EUF-CMA security proofs for novel problems can also lead to concrete breaks. In fact, Durandal [2], a promising code-base signature scheme with an EUF-CMA security proof, was recently attacked in [3]. Thus, an EUF-CMA security proof does not prevent from cryptanalysis.

Although we do not provide such a security proof, we consider attacks exploiting the leakage via published hash/signature pairs, and design our scheme integrating countermeasures for those attacks.

This paper is structured as follows: In Sect. 2, we introduce the notation that is used throughout this paper and recall the required coding-theoretic basics. In Sect. 3, we describe the proposed Lee metric signature scheme FuLeeca. In Sect. 4, we analyze the security of the proposed scheme. We first consider the best known solver to find d codewords of given Lee weight codewords, and secondly we provide heuristics for EUF-CMA security, which allow the signature scheme to be used multiple times. Finally, in Sect. 5, we analyze the performance of the scheme. We also compare the key sizes, the signature size and the computation time for signing and verification to other post-quantum signature schemes. Section 7 concludes the paper.

Remark. After this paper has been accepted, we have been noticed about an attack on the scheme.[2] Even though the attack is preliminary, i.e., there is no paper which fully describes the attack, we believe it may have an important impact on FuLeeca. We will describe the idea of the attack and possible countermeasures at the end of the paper in Sect. 6.

2 Preliminaries

2.1 Notation

Throughout this work, we denote by $\mathbb{F}_p$ the finite field of order p, where p is a prime. We often choose to represent this prime field as $\{-\frac{p-1}{2}, \dots, 0, \dots, \frac{p-1}{2}\}$, which we call the symmetric representation. We denote vectors in bold lowercase and matrices in bold uppercase letters. We refer to the i-th element of the vector $\boldsymbol{v}$ by v_i and similarly, to the j-th row of a matrix $\boldsymbol{A}$ by $\boldsymbol{a}_j$ and we denote the element in the j-th row and k-th column by $a_{j,k}$. The identity matrix of size n is denoted by $\boldsymbol{I}_n$. We denote by uppercase letters sets and for a set $S \subset \{1, \dots, n\}$, we denote by $|S|$ the cardinality and by $S^C = \{1, \dots, n\} \setminus S$ the complement. For a set $S \subset \{1, \dots, n\}$ of size s and matrix $\boldsymbol{A} \in \mathbb{F}_p^{k \times n}$, we denote by $\boldsymbol{A}_S$ the $k \times s$ matrix formed by the columns of $\boldsymbol{A}$ indexed by S, similarly for a vector $\boldsymbol{x} \in \mathbb{F}_p^n$, we denote by $\boldsymbol{x}_S$ the vector of length s formed by the entries of $\boldsymbol{x}$ indexed by S.

The sampling of an element a from the uniform distribution over a set $\mathcal{K}$ is denoted by $a \xleftarrow{\$} \mathcal{K}$. While the sampling of an element a according to a distribution χ is given by $a \xleftarrow{\$} \chi$ and by a slight abuse of notation we denote sampling of a vector $\boldsymbol{v}$ independently and identically distributed (i.i.d.) from χ by $\boldsymbol{v} \xleftarrow{\$} \chi$.

The binary entropy function with parameter p is defined as $h_2(p) := -p\log_2(p) - (1-p)\log_2(1-p)$.

[2] https://groups.google.com/a/list.nist.gov/g/pqc-forum/c/KvIege2EbuM.

2.2 Basic Cryptographic Tools

We denote the security parameter by λ. We use standard definitions of probabilistic polynomial time algorithms. We denote by "Hash" a Hash function in the perfect random oracle model.

In a digital signature scheme, we have two parties, *the signer* and the *verifier*, and three efficiently computable algorithms: the key generation, the signature generation and the signature verification. In the key generation, the signer randomly samples a secret key sk and computes and publishes the connected public key pk. For the signature generation, given a message $\boldsymbol{m}$, the signer then uses the secret key sk to compute a signature $\boldsymbol{v}$. The signer then sends $(\boldsymbol{m}, \boldsymbol{v})$ to the verifier. The verifier checks the validity of the signature $\boldsymbol{v}$ for the message $\boldsymbol{m}$ under the constraints imposed by the scheme using the public key in the signature verification step. An adversary might try to construct a valid signature, either using just the knowledge of the public key or after having observed several signatures corresponding to different messages. The adversary should only succeed with negligible probability, e.g., $< 2^{-\lambda}$.

2.3 Lee-Metric Codes

An $[n, k]$ *linear code* $\mathcal{C}$ is a k-dimensional linear subspace of $\mathbb{F}_p^n$ and can be compactly represented either through a *generator matrix* $\boldsymbol{G} \in \mathbb{F}_p^{k \times n}$, which has the code as its image or through a *parity-check matrix* $\boldsymbol{H} \in \mathbb{F}_p^{(n-k) \times n}$ having the code as its kernel. The elements of a code are called *codewords* and for any $\boldsymbol{x} \in \mathbb{F}_p^n$, we call $\boldsymbol{s} = \boldsymbol{x}\boldsymbol{H}^\top$ the *syndrome* of $\boldsymbol{x}$. The *rate* of an $[n, k]$ code is $R = \frac{k}{n}$.

For an $[n, k]$ linear code $\mathcal{C}$ and a set $I \subset \{1, \ldots, n\}$, we denote by $\mathcal{C}_I$ the set of restrictions on codewords restricted to the coordinates specified in I. We say that $I \subset \{1, \ldots, n\}$ of size k is an *information set*, if $|\mathcal{C}_I| = |\mathcal{C}|$. As a consequence, we have that for a generator matrix $\boldsymbol{G}$, respectively a parity-check matrix $\boldsymbol{H}$ of the code, $\boldsymbol{G}_I$ and $\boldsymbol{H}_{I^C}$ are invertible. We say that a generator matrix $\boldsymbol{G}$, respectively a parity-check matrix $\boldsymbol{H}$, is in systematic form (with respect to I), if $\boldsymbol{G}_I = \boldsymbol{I}_k$, respectively $\boldsymbol{H}_{I^C} = \boldsymbol{I}_{n-k}$.

Classically, we endow the vector space $\mathbb{F}_p^n$ with the Hamming metric, where the *Hamming weight* of a vector $\boldsymbol{v}$, denoted by $\mathrm{wt}_H(\boldsymbol{v})$, is given by the number of non-zero entries of $\boldsymbol{v}$. However, for this scheme, we are interested in a different metric, called the Lee metric.

The *Lee weight* of an element $a \in \mathbb{F}_p$ is defined as

$$\mathrm{wt}_L(a) := \min\{a, p - a\}, \tag{1}$$

where the representation of a is chosen to be in $\{0, \ldots, p-1\}$. In fact, one can think of the Lee weight as the L_1-norm modulo p. Clearly, the Lee weight of an element can be at most $(p-1)/2$ and therefore, we will denote this value by M. For a vector $\boldsymbol{v} \in \mathbb{F}_p^n$, its Lee weight is defined as the sum of the Lee weights of its elements, i.e.,

$$\mathrm{wt}_L(\boldsymbol{v}) := \sum_{i=1}^{n} \mathrm{wt}_L(v_i). \tag{2}$$

Note that, $\text{wt}_H(\boldsymbol{v}) \leq \text{wt}_L(\boldsymbol{v}) \leq M\text{wt}_H(\boldsymbol{v})$ and the average Lee weight of the vectors in $\mathbb{F}_p^n$ is given by $(M/2)n$.

The Lee weight induces the *Lee distance*, which we define by $d_L(\boldsymbol{x}, \boldsymbol{y}) := \text{wt}_L(\boldsymbol{x} - \boldsymbol{y})$, for all $\boldsymbol{x}, \boldsymbol{y} \in \mathbb{F}_p^n$. For a linear code $\mathcal{C}$ we define the *minimum Lee distance* as

$$d_L(\mathcal{C}) = \min\{\text{wt}_L(\boldsymbol{c}) \mid \boldsymbol{c} \in \mathcal{C}, \boldsymbol{c} \neq 0\}.$$

We denote by δ the relative minimum Lee distance, that is $\delta = \frac{d_L(\mathcal{C})}{nM}$. Let us denote by $V_L(p, n, r)$ the *Lee sphere* of radius t

$$V_L(p, n, t) := \{\boldsymbol{x} \in \mathbb{F}_p^n \mid \text{wt}_L(\boldsymbol{x}) = t\},$$

and by

$$F_L(p, T) = \lim_{n \to \infty} \frac{1}{n} \log_p(|V_L(p, n, TnM)|)$$

its asymptotic size. The exact formulas for the size of $V_L(p, n, t)$ and $F_L(p, T)$ can be found in [37,56].

Let us denote by $A(n, \delta)$ the maximal size of a code in $\mathbb{F}_p^n$ of minimum Lee distance δMn and by

$$R(\delta) = \limsup_{n \to \infty} \frac{1}{n} \log_p(A(n, \delta)).$$

The Gilbert-Varshamov (GV) bound in the Lee-metric [4] then states:

$$R(\delta) \geq 1 - F_L(p, \delta).$$

In [21], it was shown that random Lee-metric codes attain with high probability the Lee-metric GV bound, i.e., a random code has with high probability a relative minimum Lee distance δ such that $R(\delta) = 1 - F_L(p, \delta)$. For the considered quasi-cyclic code of rate 1/2, the corresponding minimum Lee distance δ of codes on the GV bound will only depend on p and is thus denoted by δ_p^{GV}.

If $\mathcal{C} \in \mathbb{F}_p^n$ is a random code of dimension k, we can also compute the expected number of codewords of a given Lee weight w as

$$|V_L(p, n, w)| p^{k-n}.$$

3 System Description

In this section, we describe how FuLeeca works.

For our scheme we represent the elements of $\mathbb{F}_p$ as

$$\left\{-\frac{p-1}{2}, \ldots, 0, \ldots, \frac{p-1}{2}\right\}$$

for $p > 3$ prime and $n \in \mathbb{N}$ even. As usual, we write M for the maximal Lee weight in $\mathbb{F}_p$, that is $M = \frac{p-1}{2}$. We define a function $\text{sgn}(x)$, that gives us the sign of an element in $\mathbb{F}_p$.

Definition 1 (Signum). *For* $x \in \mathbb{F}_p = \{-\frac{p-1}{2}, \dots, 0, \dots, \frac{p-1}{2}\}$ *let*

$$\operatorname{sgn}(x) = \begin{cases} 0 & \text{if } x = 0, \\ 1 & \text{if } x > 0, \\ -1 & \text{if } x < 0. \end{cases}$$

For the symmetric representation of $\mathbb{F}_p$, this corresponds to the common signum function.

Furthermore, we define a matching function $\operatorname{mt}(\boldsymbol{x}, \boldsymbol{y})$ that compares $\boldsymbol{x}$ and $\boldsymbol{y}$ and counts the number of symbols that hold the same sign.

Definition 2 (Sign Matches). *Let* $\boldsymbol{x}, \boldsymbol{y} \in \mathbb{F}_p^n$ *and consider the number of matches in their sign such that*

$$\operatorname{mt}(\boldsymbol{x}, \boldsymbol{y}) = \left| \{ i \in \{1, \dots, n\} \mid \operatorname{sgn}(x_i) = \operatorname{sgn}(y_i), x_i \neq 0, y_i \neq 0 \} \right|.$$

We are interested in upper bounding the probability of an attacker being able to reuse any of the previously published signatures. For that, we introduce a function calculating the probability that a vector and a uniformly random hash digest (in $\{\pm 1\}^n$) have μ sign matches. When talking about the security of the signature scheme, we will usually consider the negative $\log_2$ of this probability.

Definition 3 (Logarithmic Matching Probability (LMP)). *For a fixed* $\boldsymbol{v} \in \mathbb{F}_p^n$ *and* $\boldsymbol{y} \xleftarrow{\$} \{\pm 1\}^n$, *the probability of* $\boldsymbol{y}$ *to have* $\mu := \operatorname{mt}(\boldsymbol{y}, \boldsymbol{v})$ *sign matches with* $\boldsymbol{v}$ *is*

$$B(\mu, \operatorname{wt}_H(\boldsymbol{v}), 1/2),$$

where $B(k, n, q)$ *is the binomial distribution defined as*

$$B(k, n, q) = \binom{n}{k} q^k (1 - q)^{n-k} .$$

To ease notation, we write $\operatorname{LMP}(\boldsymbol{v}, \boldsymbol{y}) = -\log_2(B(\mu, \operatorname{wt}_H(\boldsymbol{v}), 1/2))$.

Note that this function can be efficiently approximated via additions and subtractions of precomputed values of $\log_2(x!)$, i.e., using a look-up table.

In [11], the authors computed the marginal distribution of entries where vectors are uniformly distributed in $V_L(p, n, w)$. Let E denote a random variable corresponding to the realization of an entry of $\boldsymbol{x} \in \mathbb{F}_p^n$. As n tends to infinity, we have the following result on the distribution of the elements in $\boldsymbol{x} \in \mathbb{F}_p^n$.

Lemma 4 ([11, **Lemma 1**]). *For any* $x \in \mathbb{F}_p$, *the probability that one entry of* $\boldsymbol{x}$ *is equal to* x *is given by*

$$p_w(x) = \frac{1}{Z(\beta)} \exp(-\beta \operatorname{wt}_L(x)),$$

where $Z(\beta) = \sum_{i=0}^{p-1} \exp(-\beta \operatorname{wt}_L(x))$ *denotes the normalization constant and* β *is the unique solution to* $w = \sum_{i=0}^{p-1} \operatorname{wt}_L(i) p_w(x)$.

Definition 5 (Typical Lee Set). *For a fixed weight w, let $p_w(x)$ be the probability from Lemma 4 of the element $x \in \mathbb{F}_p$. Then, we define the typical Lee set as*

$$T(p, n, w) = \left\{ \boldsymbol{x} \in \mathbb{F}_p^n \mid \boldsymbol{x}_i = x \text{ for } f(p_w(x)n) \text{ coordinates } i \in \{1, \ldots, n\} \right\},$$

for a rounding function f. That is the set of vectors, for which the element x occurs $f(p_w(x)n)$ times.

In principle, f could be simply chosen as the rounding function. This would, however, mean that the elements of $T(p, n, w)$ do not have Lee weight w in general. This effect is particularly evident when moderate values w are picked, for which the number of occurrences would be rounded to zero for many field elements.

Therefore, to obtain a closer approximation of the target weight, we design f as follows: if the expected number of occurrences for a symbol $x \in \mathbb{F}_p$ according to $p_w(x)n$ is at least 1, we always round down. If, however, the element x is expected to occur at most once, we round up or down according to a threshold τ. This τ allows us fine control over the Lee weight of the vector $\boldsymbol{x} \in T(p, n, w) \subset \mathbb{F}_p^n$. We choose this value such that the vector used to generate the secret key has a Lee weight as close to the GV bound as possible.

3.1 Key Generation

The key generation of our signature scheme is presented in Algorithm 1. The basic idea to generate the secret key $\boldsymbol{G}_{sec}$ is to sample two cyclic matrices $\boldsymbol{A}, \boldsymbol{B} \in \mathbb{F}_p^{n/2 \times n/2}$ of Lee weight $w_{key} = \delta_p^{\mathrm{GV}} n$, where $\boldsymbol{A}$ has to fulfill the extra property of being an invertible matrix. Note that this property is satisfied for random matrices with large probability. However, as $\boldsymbol{A}$ is a cyclic matrix of $\boldsymbol{a}$ sampled from the typical Lee set and symmetric in 0, $\boldsymbol{A}$ is not invertible as is. Therefore, the sign of each coefficient in $\boldsymbol{a}$ is randomly flipped such that $\boldsymbol{a}$ is not symmetric in 0 anymore. The public key is obtained by computing the row reduced Echelon form of $\boldsymbol{G}_{sec}$, referred to as $\boldsymbol{G}_{sys}$. The public key is then formed by the non-trivial part of $\boldsymbol{G}_{sys}$, which we denote by $\boldsymbol{T}$.

Algorithm 1: Key Generation

Input: Prime p, code length n, security level λ, Lee weight w_{key}

1 Choose $\boldsymbol{a}, \boldsymbol{b} \xleftarrow{\$} T(p, n/2, w_{key}/2)$.
2 Randomly flip each coefficient's sign in $\boldsymbol{a}$
3 Construct cyclic matrix $\boldsymbol{A} \in \mathbb{F}_p^{n/2 \times n/2}$ from all shifts of $\boldsymbol{a}$. $\boldsymbol{A}$ needs to be invertible. If this is not the case, resample $\boldsymbol{a}$ according to Line 1.
4 Construct cyclic matrix $\boldsymbol{B} \in \mathbb{F}_p^{n/2 \times n/2}$ from all shifts of $\boldsymbol{b}$.
5 Generate the secret key $\boldsymbol{G}_{\text{sec}} = \begin{pmatrix} \boldsymbol{A} & \boldsymbol{B} \end{pmatrix} \in \mathbb{F}_p^{n/2 \times n}$.
6 Calculate the systematic form $\boldsymbol{G}_{\text{sys}} = \begin{pmatrix} \boldsymbol{I}_{n/2} & \boldsymbol{T} \end{pmatrix}$ of $\boldsymbol{G}_{\text{sec}}$ with $\boldsymbol{T} = \boldsymbol{A}^{-1}\boldsymbol{B}$.

Output: public key $\boldsymbol{T}$, private key $\boldsymbol{G}_{\text{sec}}$

Note that $|T(p, n/2, w_{key})|^2$ corresponds to the cardinality of our key space. In order to prevent brute force attacks, this cardinality needs to be larger than 2^λ.

3.2 Signature Generation

Note that most of the Hash-and-Sign schemes require the Hash of a message to be a syndrome for a public parity-check matrix. In this Hash-and-Sign algorithm, we proceed differently. We use the generator matrix to generate signatures which are codewords of Lee weight within a fixed range. The connection to the Hash of the message vector is established through the number of sign matches.

The signature generation takes as its input the message $\boldsymbol{m}$ to be signed and makes use of the private key $\boldsymbol{G}_{sec}$ and outputs the signature $\boldsymbol{y}$. To do so the algorithm utilizes the secret generators matrix of the code, namely the rows of $\boldsymbol{G}_{sec}$, to find a codeword $\boldsymbol{v} = [\boldsymbol{y}, \boldsymbol{yT}]$ of Lee weight in $[w_{sig} - \varepsilon_s, w_{sig}]$ with sign matches achieving a desired LMP between the hash of the message and the signature codeword. Without having access to a secret basis (the private key), it is already computationally hard to find codewords in the desired Lee weight range (even ignoring the LMP). Therefore, this property suffices to ensure that it is hard to generate fresh codewords that can function as signatures even for arbitrary hashes.

Loosely speaking, a high LMP value ensures that enough signs of the codeword $\boldsymbol{v}$ and challenge $\boldsymbol{c}$ match. This establishes the connection between the signature and the message and prevents reusing codewords contained in previously published signatures to sign freshly generated hashes. Sampling a fresh salt if a signing attempt does not work guarantees that any message can be signed successfully.

In line 1, one takes the secret key $\boldsymbol{G}_{sec}$ from the Key Generation 1 and stacks it with its negative $-\boldsymbol{G}_{sec}$. In line 2, we hash the input message and get $\boldsymbol{m}'$, which will be fed together with a salt to CSPRNG in line 5 to get the target vector $\boldsymbol{c}$ for the number of sign matches, i.e., the LMP$(\boldsymbol{v}, \boldsymbol{c})$, where $\boldsymbol{v}$ denotes the information vector of the signature $\boldsymbol{y}$. Line 6 assures that $\boldsymbol{c}$ is in $\{\pm 1\}^n$ making its signs comparable with the signs of vectors in $\mathbb{F}_p^n$. In line 9, we are checking how many matches the row $\boldsymbol{g}_i$ has with the target vector $\boldsymbol{c}$. We take into account how many of the signs of $\boldsymbol{c}$ and $\boldsymbol{g}_i$ are matching in line 10, where $\lfloor \cdot \rfloor$ denotes truncation. We do this by setting the magnitude in the corresponding position of the information vector according to the number of matches and the scaling factor s. Thus, if the row has many matches with the target $\boldsymbol{c}$, we add this row multiple times. This results in the information vector $\boldsymbol{x}$ and in line 12 produces the preliminary codeword $\boldsymbol{v}$.

Lines 11–33, which we refer to as the *Concentrating* procedure, are necessary to ensure that the signatures vary as little as possible in Lee weight and sign matches.

$\mathcal{A}$ keeps track of which rows have already been added or subtracted from the codeword $\boldsymbol{v}$ and is updated respectively in line 26, 28. In line 14, we initiate the condition lf with 1, which keeps track of whether the conditions of the signature

Algorithm 2: Signing

Input: Secret key $\boldsymbol{a}, \boldsymbol{b}$, message $\boldsymbol{m}$, threshold ε, signature weight w_{sig}, key weight w_{key}, scaling factor $s \in \mathbb{R}$, security level λ, number of concentrating iterations n_{con}.

Output: salt, signature $\boldsymbol{y}$.

1 $\boldsymbol{G}_{sec} \leftarrow (\boldsymbol{A}, \boldsymbol{B})$, $\boldsymbol{G} = \begin{pmatrix} \boldsymbol{G}_{sec} \\ -\boldsymbol{G}_{sec} \end{pmatrix}$ with rows $\boldsymbol{g}_i'$

2 $\boldsymbol{m}' \leftarrow \text{Hash}(\boldsymbol{m})$

3 **repeat**

4   salt $\xleftarrow{\$} \{0,1\}^{256}$ `// Simple signing starts`

5    $\boldsymbol{c} \leftarrow \text{CSPRNG}(\boldsymbol{m}' \,||\, \text{salt})$

6    $c_i \leftarrow (-1)^{c_i} \quad \forall i$

7    $\boldsymbol{x} \leftarrow (0, \ldots, 0)$

8   **for** $i \leftarrow 1$ **to** $n/2$ **do**

9      $x_{mt} = \text{mt}(\boldsymbol{g}_i, \boldsymbol{c}) - \frac{\text{wt}_H(\boldsymbol{g}_i)}{2}$

10      $\boldsymbol{x}_i = \lfloor x_{mt} s \rfloor$ `// Simple signing ends`

   end

11    $\mathcal{A} \leftarrow \{1, \ldots, n\}$ `// Allowed row index set`

12    $\boldsymbol{\nu} \leftarrow \boldsymbol{x}\boldsymbol{G}_{sec}$ `// Concentrating starts`

13    $\boldsymbol{\nu}' \leftarrow (0, ..., 0)$, $i' = 0$

14    $lf \leftarrow 1$

15   **for** $j \leftarrow 1$ **to** n_{con} **do**

16   **for** $i \in \{1, \ldots, n\}$ **do**

17        $\boldsymbol{\nu}'' \leftarrow \boldsymbol{\nu} + \boldsymbol{g}_i'$

18   **if** $|\text{LMP}(\boldsymbol{\nu}'', \boldsymbol{c}) - (\lambda + 64 + \varepsilon)| \leq |\text{LMP}(\boldsymbol{\nu}', \boldsymbol{c}) - (\lambda + 64 + \varepsilon)|$ **then**

19   **if** $i \in \mathcal{A} \,||\, lf = 0$ **then**

20            $\boldsymbol{\nu}' \leftarrow \boldsymbol{\nu}''$, $i' \leftarrow i$

     end

21      $w' \leftarrow \text{wt}_L(\boldsymbol{\nu}')$

22   **if** $w' > w_{sig} - w_{key}$ **then**

       $lf \leftarrow 0$

23   **if** $w' \leq w_{sig}$ **then**

24        $\boldsymbol{\nu} \leftarrow \boldsymbol{\nu}'$

25   **if** $i' \leq \frac{n}{2}$ **then**

26        $\mathcal{A} \leftarrow \mathcal{A} \setminus \{i' + n/2\}$

27   **else**

28        $\mathcal{A} \leftarrow \mathcal{A} \setminus \{i' - n/2\}$

   end

29   **if** $\text{wt}_L(\boldsymbol{\nu}) \leq w_{sig}$ && $\text{wt}_L(\boldsymbol{\nu}) > w_{sig} - 2w_{key}$ && $\text{LMP}(\boldsymbol{\nu}, \boldsymbol{c}) \geq \lambda + 64$ **then**

30      $[\boldsymbol{y}, \boldsymbol{y}\boldsymbol{T}] \leftarrow \boldsymbol{\nu}$

31   **return** salt, ENCODE($\boldsymbol{y}$)

32   **else**

33   go to Line 3 `// Concentrating ends`

end

(that is LMP and Lee weight) are satisfied, in which case lf will be set to 0. To ensure a constant time signature generation, the lines 16–28 will only run up to n_{con} times.

To have signatures with much lower Lee weight than other signatures is undesirable, as this might leak information on the secret key. Thus, the iterative approach in lines 16–20 is used to add or subtract the generator row minimizing the absolute difference to the desired LMP. For this, we first add the row $\boldsymbol{g}'_i$ to $\boldsymbol{v}$ in line 17 and then check in line 18 if the difference of the LMP to the target is minimized by adding this row. Line 19 checks whether the row $\boldsymbol{g}'_i$ is within the set of allowed rows, i.e., in $\mathcal{A}$, or if the signature conditions are satisfied, i.e., $lf = 0$. This results in a codeword $\boldsymbol{v}'$, which is close enough to the target LMP.

Lines 21–24 aim at creating signatures of almost constant Lee weight. For this, we compute in line 21 the Lee weight w' of $\boldsymbol{v}'$ and check in line 22 if it is close enough to the target Lee weight w_{sig}, i.e., at most has a w_{key} difference. In this case, we update the signature condition lf with 0. If the Lee weight w' is larger than the target, we reset $\boldsymbol{v}'$ with the initial $\boldsymbol{v}$ in lines 23, 24. Lines 25–28 update the set of rows that are allowed to be added. In fact, if $i' \leq n/2$, we added a row of $\boldsymbol{G}_{sec}$ and excluded the same row to be extracted again by excluding $i' + n/2$ from the allowed set $\mathcal{A}$. If $i' > n/2$, the added row was from $-\boldsymbol{G}_{sec}$ and we exclude $i' - n/2$ from $\mathcal{A}$ to avoid subtracting the same row again.

After all iterations have been completed, lines 29–33 check whether the resulting codeword is within the desired LMP/Lee weight range. If this is the case, we extract the information vector $\boldsymbol{y}$ from $\boldsymbol{v}$ in line 30 and publish the signature (salt, $\text{ENCODE}(\boldsymbol{y})$). The encoding procedure $\text{ENCODE}(\cdot)$ is described in Sect. 3.4. Otherwise, another salt is sampled and the signing procedure restarts.

The scaling parameter s used in line 10 is experimentally determined with the goal of minimizing the running time of the Signing algorithm. Its value is a trade-off between the probability of creating a valid signature for a specific hash value and the amount of iterations within the *Concentrating* procedure.

3.3 Signature Verification

The verification process is quite simple. In a first step, the received signature $\boldsymbol{y}'$ is decoded as explained in Sect. 3.4 to obtain the uncompressed vector $\boldsymbol{y}$. The verifier computes in lines 3 and 4 $\boldsymbol{c}$ as CSPRNG from the hash of the message and salt. Then, the verifier checks that $\boldsymbol{v}$ is indeed a codeword of the public code; this is ensured by computing $\boldsymbol{v}$ as $[\boldsymbol{y}\ \boldsymbol{yT}]$ in line 5.

Then, the verifier checks in line 6 that the codeword $\boldsymbol{v}$ has Lee weight of at most w_{sig}. Finally, one checks whether a sufficient amount of the signs of the signature $\boldsymbol{v}$ match the output $\boldsymbol{c}$ of the CSPRNG(Hash($\boldsymbol{m}$)||salt), i.e., $\mathrm{LMP}(\boldsymbol{v}, \boldsymbol{c}) \geq \lambda + 64$. This verification process is given in Algorithm 3.

Algorithm 3: Verification

Input: signature (salt, $\boldsymbol{y}'$) message **m**, public key $\boldsymbol{T}$, Lee weight w_{sig}.

$\boldsymbol{y} \leftarrow \text{DECODE}(\boldsymbol{y}')$
$\boldsymbol{m}' \leftarrow \text{Hash}(\boldsymbol{m})$
$\boldsymbol{c} \leftarrow \text{CSPRNG}(\boldsymbol{m}' \mid\mid \text{salt})$
$c_i \leftarrow (-1)^{c_i} \quad \forall i$
$\boldsymbol{v} = [\boldsymbol{y}\ \boldsymbol{y}\boldsymbol{T}]$.
Accept if the following two conditions are satisfied:
(a) $\mathrm{wt}_L(\boldsymbol{v}) \leq w_{\text{sig}}$,
(b) $\mathrm{LMP}(\boldsymbol{v}, \boldsymbol{c}) \geq \lambda + 64$.

Otherwise, Reject.

Output: Accept or Reject

3.4 Encoding and Decoding

The coefficients that constitute a signature before encoding follow a Gaussian-like distribution centered at zero. This allows to reduce the signature size by compressing the signature and encoding it in a bitstring. For that, we use the same approach as proposed in the Falcon signature scheme [36]. That is, each coefficient is converted into its signed representation and split into a *tail* and *head.* The coefficient's sign bit is concatenated with the uncoded tail, as this tail is approximately uniformly distributed and thus cannot be compressed efficiently. The remaining bits in the coefficients head are then encoded in a 0^k1 fashion, that is a sequence of k zeroes and a one, where k is the value of the head.

4 Security Analysis

In this section, we assess the security of FuLeeca. The analysis consists of three parts. We begin by considering the generic solvers for finding codewords of given Lee weight. The second part describes known attacks and our countermeasures. The third part discusses the applicability of lattice reduction algorithms to solve the hard computational problems underlying this system. Considering all mentioned attacks, we determine the presented parameters to achieve the security levels required by NIST.

4.1 Hardness of Underlying Problem and Generic Solvers

The adversary can attempt to recover the secret key from the public key, which is known as a key recovery attack. For FuLeeca, this is equivalent to finding any of the rows of the secret generator matrix, which are of weight w_{key}. Alternatively, the attacker can try to forge a signature directly without knowledge of the secret key. Forging a signature of FuLeeca is, therefore, equivalent to finding a codeword of given Lee weight that satisfies both the number of required matches and the weight restriction.

Hence, both attacks require solving instances of the finding a codeword of given Lee weight problem, which is formally defined as follows.

Problem 6 (Finding Codeword of Given Lee Weight). Given $\boldsymbol{H} \in \mathbb{F}_p^{(n-k)\times n}$ and $w \in \mathbb{N}$ find a $\boldsymbol{c} \in \mathbb{F}_p^n$ such that $\boldsymbol{c}\boldsymbol{H}^\top = \boldsymbol{0}$ and $\mathrm{wt}_L(\boldsymbol{c}) = w$.

This problem has first been studied in [42]. Problem 6, i.e., finding codewords of given weight is equivalent to the decoding problem. The decisional version of this problem has been proven to be NP-complete in [56].

Several algorithms have been proposed to solve this problem, they all belong to the family of Information Set Decoding (ISD) algorithms.

Remark 7. Note that ISD algorithms can be formulated such that they solve the syndrome decoding problem, that is: given a parity-check matrix $\boldsymbol{H} \in \mathbb{F}_p^{(n-k)\times n}$, a syndrome $\boldsymbol{s} \in \mathbb{F}_p^{n-k}$ and a target weight t, they find an error vector $\boldsymbol{e} \in \mathbb{F}_p^n$, such that $\boldsymbol{H}\boldsymbol{e}^\top = \boldsymbol{s}^\top$ and $\mathrm{wt}(\boldsymbol{e}) = t$. Thus, by setting $\boldsymbol{s} = \boldsymbol{0}$, we can use such solvers to find codewords of weight t. However, note that Prange's algorithm [49] searches for a transformed syndrome $\boldsymbol{s}' = \boldsymbol{s}\boldsymbol{U}$, for some invertible $\boldsymbol{U}$ and wants the transformed syndrome to have weight t. As this is never satisfied for $\boldsymbol{s} = \boldsymbol{0}$, Prange cannot be used to find codewords of given weight. However, all improvements upon Prange, such as Stern/Dumer [29,55], MMT [45], BJMM [13] try to first enumerate the error vector in the information set and then check whether the remaining vector has the remaining weight. This can also be applied to $\boldsymbol{s} = \boldsymbol{0}$.

ISD algorithms make use of an information set of the code, where one assumes a small weight and thus constructs lists of these partial solutions.

Let us quickly recall the main steps of an ISD algorithm. Given $\boldsymbol{H} \in \mathbb{F}_p^{(n-k)\times n}$, choose an information set I and bring $\boldsymbol{H}$ into a partial systematic form. For this, let J be a set of size $k + \ell$, which contains the information set I and transform $\boldsymbol{H}$ as

$$\boldsymbol{U}\boldsymbol{H}\boldsymbol{P} = \tilde{\boldsymbol{H}} = \begin{pmatrix} \boldsymbol{I}_{n-k-\ell} & \boldsymbol{H}_1 \\ 0 & \boldsymbol{H}_2 \end{pmatrix},$$

where $\boldsymbol{U} \in \mathbb{F}_p^{(n-k)\times(n-k)}$ is an invertible matrix and $\boldsymbol{P} \in \mathbb{F}_p^{n\times n}$ is a permutation matrix. Thus, we also split the unknown solution $\boldsymbol{c}$ into the indices J and J^C, i.e., $\boldsymbol{c}\boldsymbol{P}^\top = (\boldsymbol{c}_1, \boldsymbol{c}_2)$. Assuming that $\boldsymbol{c}_2$ has Lee weight v, we get the following two equations:

$$\begin{aligned} \boldsymbol{c}_1 + \boldsymbol{c}_2\boldsymbol{H}_1^\top &= 0 \\ \boldsymbol{c}_2\boldsymbol{H}_2^\top &= 0. \end{aligned}$$

Thus, we can first solve the second equation, $c_2 H_2^\top = 0$ with $\mathrm{wt}_L(c_2) = v$ as we then can easily check if the missing part c_1 has the remaining Lee weight, by $\mathrm{wt}_L(c_2 H_1^\top) = w - v$.

In [56], several algorithms have been presented to solve the smaller instance, namely using Wagner's approach of a set partitioning and using representation technique. In [22], the authors presented the amortized Wagner's approach.

Finally, in [12], the authors presented an adaption of these algorithms, taking into account that a random low Lee weight codeword has the exponential weight distribution observed in [11]. In these papers, it has been observed that the amortized BJMM approach attains the lowest computational cost, and thus we consider this algorithm to compute the security level of the proposed parameters.

For the details of the algorithm, we refer to [12]. Mathematica programs to compute the computational costs of BJMM are publicly available[3] or for Wagner's cost here[4].

We adapted the program, which computes the classical asymptotic cost c in the form $2^{c \cdot n}$, by considering the cost $c/2$ on a capable quantum computer (see [16,22]).

Since we sample the secret vectors for the generator matrix using the typical Lee sets, i.e., any $x \in \mathbb{F}_p$ occurs in the sought-after codeword $f(p_{w_{key}}(x) \cdot n)$ number of times, it makes sense to use this information in an ISD algorithm. However, as shown in [12], the amortized BJMM algorithm outperforms even the attempts to use restricted balls in the case where we are beyond the unique decoding radius. Thus, we build our security analysis on this fastest known algorithm, taking into account also polynomial speedups due to the quasi-cyclic structure [53].

4.2 Analysis of the Algorithm with Respect to Known Attacks

We assume that an attacker has access to up to 2^{64} signatures for chosen messages. Such multi-use scenarios require an existential unforgeability under chosen message attack (EUF-CMA) security proof. For Hash-and-Sign approaches, EUF-CMA security proofs are notoriously difficult. Unfortunately, we cannot provide one at the moment. We prevented possible leakages and vulnerabilities via the *Concentrating* procedure. These considerations are described in more detail below. Note that the *Concentrating* procedure at the moment does not involve a threshold on how close the valid signatures have to be to the target LMP. This flexibility might be of use in a future EUF-CMA security proof. Additionally, the scheme does not involve rejection sampling, which might be helpful to strengthen security, as soon as new attack vectors are known.

Exploiting additional knowledge given to the attacker in the form of signatures is perhaps the most common way to attack Hash-and-Sign based signature schemes. In fact, information leaked by the signatures has repeatedly been used

---

[3] https://git.math.uzh.ch/isd/lee-isd/lee-isd-algorithm-complexities/-/blob/master/Lee-ISD-restricted.nb.

[4] https://github.com/setinski/Information-Set-Decoding-Analysis.

to retrieve the private key. To give an example, successful attacks on the schemes [6,47] have been presented in [27,51]. Specifically, these attacks exploit the fact that for the proposed schemes in the Hamming metric a basis vector as well as the signatures have low weight, i.e., a small support. The main problem in the design of these attacked schemes was that the supports of the published signatures correlate with the private key. We consider attacks exploiting leakage via published hash/signature pairs. Such support-based attacks cannot directly be applied to FuLeeca as in the Lee metric vectors of low Lee weight do not necessarily have a small Hamming support. In fact, by putting the weight of the secret generators on the GV bound, we may even treat the resulting code as a random code. This thwarts Hamming-metric attacks as the secret generators and the signatures have close to full Hamming weight.

Setting a sufficiently high threshold for the number of required sign matches prevents that a previously published signature can be directly used to sign another message. An obvious generalization of this reuse attack is creating linear combinations of existing signatures to forge new signatures. Note, however, that with overwhelming probability, the Lee weight of the resulting vector will be too large to be accepted by the verifier. Hence, such an attack, which is similar to performing a sieving algorithm known from lattice-based cryptography, requires complexity which is exponential in the code parameters. Notably, the works [38,44] show that finding a codeword of lower Lee weight in a quasi-cyclic code is significantly easier in case the code dimension $n/2$ is a composite number. In fact, the security reduces to the codeword finding problem in a quasi-cyclic code with dimension equal to the smallest factor of $n/2$. Therefore, for all considered parameter sets in this work, we choose $n/2$ to be prime.

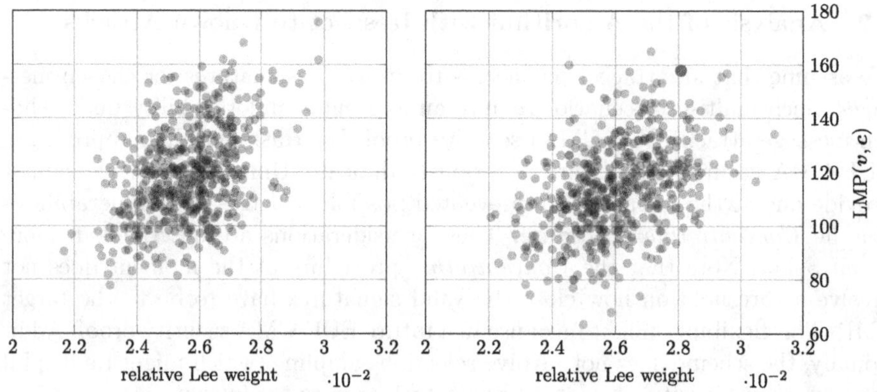

Fig. 1. Evaluation of 500 signatures for simulated hashes (i.i.d uniform) using *two different keys* (left and right) after application of "Simple Signing".

To avoid leakage via published hash/signature pairs, we integrated a specific procedure into the signing algorithm, which we refer to as the *Concentrating* procedure. In the following, we first examine the signing algorithm without applying

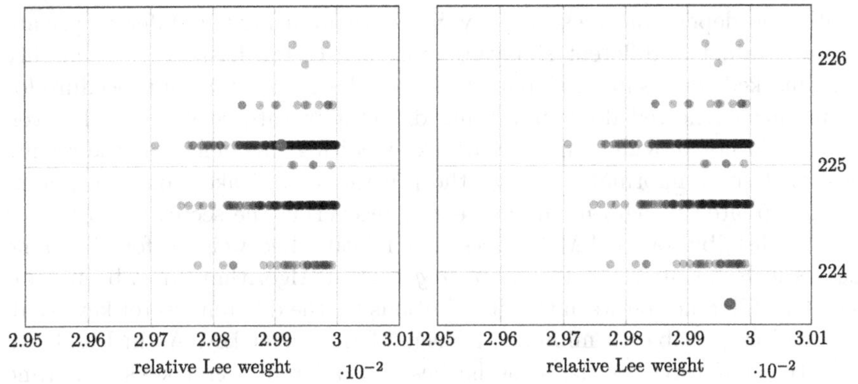

Fig. 2. Evaluation of 500 signatures for simulated hashes (i.i.d uniform) for *two different keys* after application of both "Simple Signing" and "Concentrating".

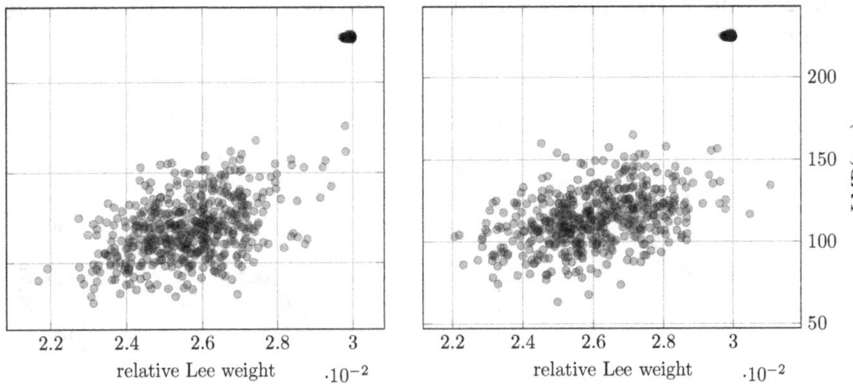

Fig. 3. Evaluation of 500 signatures for simulated hashes (i.i.d uniform) using *two different keys* both after application of the "Simple Signing" part of Algorithm 2 and as well as applying the "Concentrating" procedure (dense clusters in the upper right).

the specified *Concentrating* procedure. We randomly draw $k = 500$ salts and messages and observe the corresponding outputs of the hash-function $h_1, \ldots, h_k$, i.e., $h_\ell = \text{Hash}(\text{salt}||m_\ell)$. For two different private keys we compare the Lee weights and sign matches of the corresponding signatures after just applying "Simple Signing". Figure 1 shows the relation between the relative Lee weights and the LMP between the codeword and the target vector, which is the hash of the message. Since the signature algorithm effectively correlates the secret key and the hashes it appears to be possible to learn at least some information about the secret key based on the distribution of resulting codewords in this Lee weight / LMP space.

The distribution of signatures for both private keys of Fig. 1 show that the LMP between hash and codeword, as well as the resulting Lee weights vary significantly and depend on the secret key. Since we are using two different private keys, we obtain two different signatures for each of the hashes. To exemplify this, we marked the resulting signatures before the *Concentrating* procedure for the same hash (the red dots) but using different private keys in Fig. 1. Even though we do not provide a specific attack exploiting this behavior, the results suggest that some information about the private key is leaked and can potentially be exploited to help in the process of recovering the secret key. Figure 2 shows the distribution of LMP values and relative Lee weights for the same hashes as in Fig. 1 after the *Concentrating* part of Algorithm 2 has been completed. The difference between the distributions for the different secret keys shall be as small as possible to minimize leakage of the secret key. As in Fig. 1, we marked the signatures for the same hashes and different secret keys, this time after the *Concentrating* procedure in Fig. 2. The results show that the *Concentrating* procedure significantly reduces the leakage observable via the relative Lee weight/LMP map. Figure 3 provides the information observable from Fig. 1 and Fig. 2 within a single plot to further illustrate the effect of the *Concentrating* procedure.

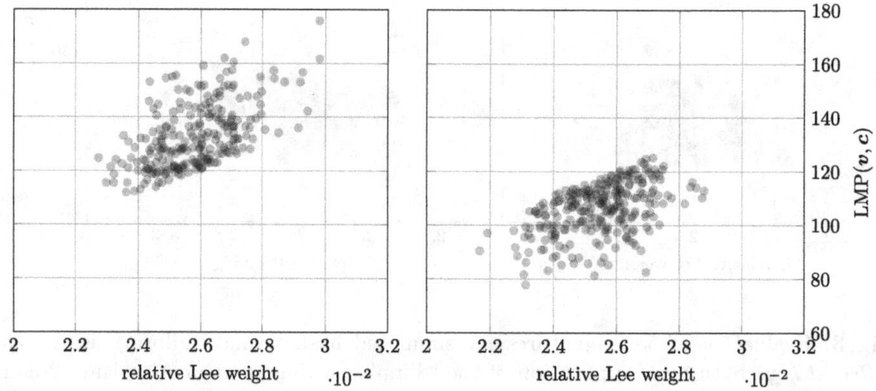

Fig. 4. Evaluation of 500 signatures for simulated hashes (i.i.d uniform) before applying the *Concentrating* procedure. Unlike the previous figures, all of the displayed signatures were created using a *single key*. The vectors are divided into two (nearly equally large) groups, where the ratio between the log probability (LMP) and the Lee weight is above average (left), respectively below average (right).

Similarly, we also observe that the shape of the distribution of signatures in the Lee weight/LMP space does not appear to meaningfully depend on the distribution of the same signatures after "Simple Signing". This is demonstrated in Fig. 4 and 5 where for a single key we apply "Simple Signing" to the same set of hashes as before but split the signatures into two groups of almost equal size. For group one (left hand side of Fig. 4), obtaining a codeword with the required LMP after application of the *Concentrating* procedure is expected to be easier

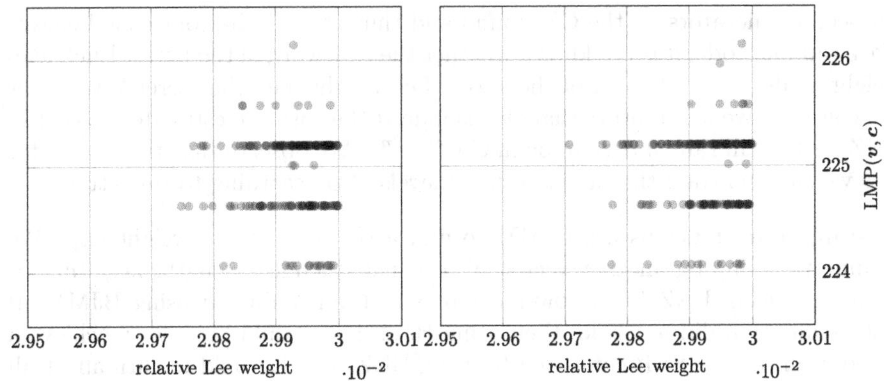

Fig. 5. The same two sets of hashes for *the same key* (as in Fig. 4) after applying the "Concentrating" algorithm.

than for group two (right hand side of Fig. 4) since in terms of the ratio between the log probability (LMP) and the Lee weight all of these are above average, while group two is below average. In fact, the percentage of hashes in group two that lead to a valid signature (right hand side of Fig. 5) in the end is slightly lower than for group one (left hand side of Fig. 5). However, this behaviour is to be expected for effectively every private key and, thus, this does not reveal any useful information about any chosen key in particular.

4.3 Lattice-Based Attacks

Since the Lee metric is close to the Euclidean metric used in lattice-based cryptography, one has to study the known combinatorial attacks therein. In fact, the Lee metric corresponds to the L_1-norm, whereas the Euclidean metric corresponds to the L_2-norm. It is well known [50] that problems with respect to the L_2-norm can be reduced to problems with respect to any other L_p-norm. This result translates to: any algorithm solving a problem in the L_p-norm can also be used to solve the problem in the L_2-norm. Or as stated in [50]: "our main result shows that for lattice problems, the L_2-norm is the easiest." Thus, one can use the Lee-metric ISD algorithms to solve lattice-based problems in the Euclidean metric. It is unknown whether the reverse direction is also possible, i.e., whether there exists a reduction from problems with respect to the L_1-norm to problems with respect to the L_2-norm. This is, however, exactly the direction required in order to use lattice-based algorithms to solve problems in the Lee metric.

To the best of our knowledge, the only sieving algorithm in the L_1-norm is provided in [20], where the authors provide an $(1+\varepsilon)$ approximation algorithm for the closest vector problem for all L_p-norms that runs in $(2+1/\varepsilon)^{\mathcal{O}(n)}$. The asymptotic cost of this algorithm does not outperform the considered Lee-metric ISD algorithms.

Another lattice-based approach is to search for the codeword of the lowest Euclidean weight, e.g., using the BKZ algorithm [52]. Since we set the weight of the secret generators on the GV bound and thus assume that our code behaves like a random code, it is not known whether the codeword of the lowest Euclidean weight is also the codeword of the lowest Lee weight, i.e., the secret key. Under the conservative assumption that this is indeed the case, we estimate the cost of BKZ for the full rank lattice to be in $\mathcal{O}\left(2^{0.292n}\right)$. We observe that the parameter sets we choose attain the target security levels also according to this attack.

Assumption 1: Let us use BJMM to find a vector $\boldsymbol{v}$ of Lee weight w_{sig}. We assume that finding another vector $\boldsymbol{v}'$ of equal Euclidean length, i.e., $||\boldsymbol{v}||_2 = ||\boldsymbol{v}'||_2$, by using BKZ has a lower complexity than finding $\boldsymbol{v}$ using BJMM. If this assumption did not hold, then using BJMM we would be able to achieve a speedup in solving SVP compared to using BKZ, which would in turn affect all lattice-based cryptosystems.

Assumption 2: We assume that the complexity of using BKZ to find a vector having Lee weight less than or equal to w_{sig} is higher compared to using BJMM for this task.

For a Lee weight of w_{sig}, the consequence of Assumption 2 not holding is that BKZ would outperform all known ISD algorithms for solving the given weight codeword finding problem at that weight. BKZ requires orthogonal projections within the LLL step. However, the L_1 norm is not induced by a scalar product and, therefore, we assume that the best way to use BKZ for finding short vectors in the L_1 norm is to use it for finding short vectors in L_2 norm and to hope that those are also short enough in the L_1 norm. We assume that using BJMM to find short vectors in the L_1 norm is more efficient than this.

5 Efficiency and Performance

5.1 Parameters

Due to the quasi-cyclic structure of the private matrix $\boldsymbol{G}_{sec}$ it is sufficient to store only one of its rows. Therefore, the size of the private key is in the order $\mathcal{O}_p(n)$, where the constant depends on the parameter p.

We take a conservative choice for the NIST security levels [43], as shown in Table 1.

Table 1. Conservative NIST Categories

NIST Security Level	Classical Cost	Quantum Cost
I	160	80
III	224	112
V	288	144

Table 2. Parameters for the proposed signature scheme FuLeeca. All sizes are given in Bytes.

NIST cat.	n	s	n_{con}	secret key size	public key size	sign. size
I	1318	0.046 875	100	2636	1318	1100
III	1982	0.035 156 25	90	3964	1982	1620
V	2638	0.023 437 5	178	5276	2638	2130

The chosen parameters and associated data sizes for the NIST categories I, III, and V are given in Table 2. An extensive comparison of FuLeeca's communication costs with other signature schemes is provided in Table 3. Note that for all parameter sets, we fix $p = 65521, \varepsilon = 1$ and the relative Lee weights $\omega_{sig} = w_{sig}/(nM) = 0.03$, and $\omega_{key} = w_{key}/(nM) = 0.001437$ is on the GV bound, where we recall that $M = \lfloor \frac{p-1}{2} \rfloor$ is the maximal Lee weight in $\mathbb{F}_p$.

The signature sizes are averaged over 1k generated compressed signatures and include the size of the salt. For compression, we have adapted the mechanisms as used in the Falcon signature scheme. Although the signature size is not constant, it can be padded to obtain a fixed size. As proposed in [30], it is possible to compress the signatures resulting from Algorithm 2 even further.

5.2 Reason for Choice of Parameters

Recall that the choice to set w_{key} on the Lee-metric GV bound is necessary to treat the public code as a random code and thus estimate the BKZ algorithms cost at $2^{0.292n}$.

We choose $p = 65\,521$ in order to set the Lee weight w_{key} of the secret generators on the Lee-metric GV bound and still have a large enough distance to the Lee weight of the signatures w_{sig}. In fact, for smaller choices of p and setting w_{key} on the Lee-metric GV bound, we cannot find enough sign matches to signatures of Lee weight w_{sig} with $w_{sig} < 0.2$. The bound $w_{sig} < 0.2$ is mandatory to avoid a polynomial time cost of ISD algorithms.

The parameters are also chosen according to the best-known attack to find a codeword of given Lee weight given our public key $\boldsymbol{G}_{pub}$, namely the quantum, amortized BJMM algorithm in the Lee metric.

For the choice of $p = 65\,521$, one cannot explicitly compute the cost of the BJMM algorithm using the program[5] due to numerical instabilities. A conservative extrapolation from results for smaller choices of p suggests that the cost for BJMM at $w_{sig} = 0.03$ lies at $2^{0.08n}$. We want to note here that Wagner's algorithm implies a cost of $2^{0.5n}$.

[5] https://git.math.uzh.ch/isd/lee-isd/lee-isd-algorithm-complexities/-/blob/master/Lee-ISD-restricted.nb.

Table 3. Comparison of post-quantum signature schemes for NIST level I (except for Dilithium which achieves NIST level II). All sizes are given in kB.

scheme	public key size	signature size	total size	variant
Falcon [36]	0.9	0.6	1.5	-
FuLeeca [This work]	1.3	1.1	2.4	-
Dilitihium [28]	1.3	2.4	3.7	-
R-BG [7]	0.1	7.7	7.8	Fast
	0.1	7.2	7.3	Short
Rank SDP Fen [32]	0.9	7.4	8.3	Fast
	0.9	5.9	6.8	Short
Ideal Rank BG [19]	0.5	8.4	8.9	Fast
	0.5	6.1	6.6	Short
PKP BG [19]	0.1	9.8	9.9	Fast
	0.1	8.8	8.9	Short
SDItH [34]	0.1	11.5	11.6	Fast
	0.1	8.3	8.4	Short
Ret. of SDitH [1]	0.1	12.1	12.1	Fast, V3
	0.1	5.7	5.8	Shortest, V3
SPHINCS$^+$ [5]	<0.1	16.7	16.7	Fast
	<0.1	7.7	7.7	Short
Beu [17]	0.1	18.4	18.5	Fast
	0.1	12.1	12.2	Short
Durandal [2]	15.2	4.1	19.3	-
FJR [33]	0.1	22.6	22.7	Fast
	0.1	16.0	16.1	Short
GPS [41]	0.1	24.0	24.1	Fast
	0.1	19.8	19.9	Short
MinRank Fen [32]	18.2	9.3	27.5	Fast
	18.2	7.1	25.3	Short
LESS-FM [8]	10.4	11.6	23.0	Balanced
	205.7	5.3	211.0	Short sign
WAVE [26]	3200	2.1	3202	-

We choose the length n according to the BKZ algorithm on full-rank lattices, which runs with a cost of $2^{0.292n}$. We aim at the conservative classical security levels $\lambda_1 = 160, \lambda_3 = 224, \lambda_5 = 288$ and set n at least such that

$$2\lambda_i + 64 = 0.292n.$$

This choice is conservative in two ways. Not only the security levels λ_i have been chosen conservatively but, also assuming a loss in the security level of λ_i

for each of the provided 2^{64} signature vectors is a very conservative approach within the estimation of the resulting security level. In fact, the parameters are chosen in such a way that even for the aforementioned loss of $\lambda_i + 64$ bits, a security level of at least λ_i bits is maintained for the respective parameter sets. It is possible to speed up solving the SVP using BKZ by providing the algorithm with short Euclidean lattice vectors [25]. The obtainable speedup is upper bounded by the cost of finding the provided lattice vectors since otherwise, we would have found an improved lattice reduction algorithm. The exact speedup obtained from integrating the short (codeword) vectors depends on their Euclidean length. However, we assume that a vector of comparable Euclidean length can be obtained at a lower cost using BKZ compared to using BJMM. We conservatively add 64 to account for the maximum possible speedup once 2^{64} signatures have been published.

In fact, we choose n even slightly larger to ensure that we reach the necessary LMP with good probability. This leads to the following lengths: $n_1 = 1318$, which ensures that $n_1/2 = 659$ is prime, $n_3 = 1982$, which ensures that $n_3/2 = 991$ and finally $n_5 = 2638$, which ensures that $n_5/2 = 1319$ is prime.

Parameter Choice I. The parameter choice $p = 65\,521$, $n = 1318$, $\omega_{sig} = w_{sig}/(nM) = 0.03$, $\omega_{key} = w_{key}/(nM) = 0.001437$ leads at least to the desired quantum cost of 2^{80}, since BJMM's algorithm indicates a quantum complexity of $2^{80} = 2^{0.08n}$ operations and the BKZ algorithm requires at least a classical complexity of $2^{384} = 2^{0.292n}$.

Parameter Choice III. The parameter choice $p = 65\,521$, $n = 1982$, $\omega_{sig} = w_{sig}/(nM) = 0.03$, $\omega_{key} = w_{key}/(nM) = 0.001437$ leads to the desired quantum cost of 2^{112}, since BJMM's algorithm indicates a quantum complexity of $2^{112} = 2^{0.08n}$ operations and the BKZ algorithm requires at least a classical complexity of $2^{578} = 2^{0.292n}$.

Parameter Choice V. The parameter choice $p = 65\,521$, $n = 2638$, $\omega_{sig} = w_{sig}/(nM) = 0.03$, $\omega_{key} = w_{key}/(nM) = 0.001437$ leads to the desired quantum cost of 2^{144}, since BJMM's algorithm indicates a quantum complexity of $2^{144} = 2^{0.08n}$ operations and the BKZ algorithm requires at least a classical complexity of $2^{770} = 2^{0.292n}$.

5.3 Detailed Performance Analysis

To evaluate FuLeeca, we provide a constant-time C reference implementation that is publicly available at https://gitlab.lrz.de/tueisec/fuleeca-signature. Both the hash functions as well as the CSPRNGs were instantiated with SHA-3 primitives. More precisely, we use the SHA-3 hash functions, as specified in FIPS 202 [54], with a digest size of 2λ for the message hashing and expand this message digest together with a salt using the eXtendable-Output Function (XOF) SHAKE256 from the FIPS 202 specification as CSPRNG.

Table 4 shows the required clock cycles and run time in milliseconds for the reference implementation of the algorithm averaged over 10 000 runs. These values were obtained on an Ubuntu 22.04 machine with an Intel Comet Lake (Intel Core i7-10700) CPU at its base frequency of 2900 MHz and 64 GB of RAM using GCC version 11.3.0 and an O3 optimization. In order to generate reliable results, all dynamic performance enhancement and power management features like hyper-threading, turbo boost, and dynamic undervolting of the CPU were disabled. Clock cycles are measured using the internal performance registers of the CPU using the library *libcpucycles*[6].

Table 4. Runtime of the constant-time reference implementation in kilocycles and milliseconds on an Intel Comet Lake with a base frequency of 2900 MHz averaged over 10000 runs.

NIST cat.	Unit	Keygen	Sign	Verify
I	kCycles	53 913	1 803 104	1452
	ms	18	621	0.49
III	kCycles	111 937	2 139 170	2534
	ms	38	737	0.86
V	kCycles	195 729	11 805 175	3845
	ms	67	4070	1.32

6 Preliminary Attack on FuLeeca

After this paper has been accepted, we have been noticed about an attack on the scheme. Even though the attack is preliminary, i.e., there is no paper that fully describes the attack, we believe it may have an important impact on FuLeeca. We now briefly describe how the attack works and possible countermeasures.

The signatures are codewords of the code generated by the secret generator matrix G_{sec}, thus, any signature is $v = xG_{sec} \mod p$. The attack experimentally observes that for the current parameters, no modular reduction is required, that is $v = xG_{sec}$, and therefore any signature v leaks the $\mathbb{Z}$-span of G_{sec}. The lattice $L(G_{sec}) \in \mathbb{Z}^n$ has rank $n/2$ as it is enough to consider the first $n/2$ coefficients of v and recover the lattice $L(G')$, where $G_{sec} = (G', G'')$.

The attack experimentally observes that the $n/2$ shortest vectors (in the Euclidean metric) of the lattice $L(G')$ are the rows of G'. Thus, using the BKZ algorithm, these secret generators can be retrieved with a cost in $\mathcal{O}\left(2^{n/4 \; 0.292}\right)$, instead of the predicted $\mathcal{O}\left(2^{n/2 \; 0.292}\right)$. Finally, since $L(G')$ is a circulant lattice, the attack predicts that the actual cost of retrieving the shortest vectors is sub-exponential.

[6] The implementation is publicly available at https://cpucycles.cr.yp.to/.

Several countermeasures can be employed, such as forcing signatures v which require modular reduction, that is $v = xG_{sec} \mod p$ but $v \neq xG_{sec}$. This countermeasure, however, is expected to lead to much larger weights of the signatures, which will facilitate forgery attacks. One could use a generator matrix $G \in \mathbb{F}_p^{k \times n}$ which is not quasi-cyclic and thus will not lead to a circulant lattice $L(G')$ and additionally increase the length n, to cope with the rank k of the lattice. In this case, however, the public key size will increase drastically to $(n-k)k\log_2(p)$. For the NIST category I and $k = 1318 = n/2$, the public key size is roughly 3.5 MB and thus impractical.

7 Conclusion

In this paper, we proposed a Hash-and-Sign signature scheme based on the hardness of finding a codeword of given Lee weight. Taking known statistical attacks into account, we refined the simple signing process to render the scheme multiple-use. We keep the EUF-CMA security proof as an open problem. The scheme can be efficiently implemented as it only uses simple arithmetics and is able to achieve short signatures of 1100 bytes and public keys of 1318 bytes for the NIST category I security level. This compares favorably to the state-of-the-art of lattice-based and code-based schemes. Unfortunately, due to the attack mentioned in Sect. 6, we do not recommend the use of FuLeeca in its current version.

Acknowledgments. We would like to thank Sabine Pircher, Georg Sigl, Thomas Debris-Alazard and Wessel van Woerden for meaningful discussions.

Violetta Weger is supported by the European Union's Horizon 2020 research and innovation program under the Marie Skłodowska-Curie grant agreement no. 899987. Sebastian Bitzer, Georg Maringer, Stefan Ritterhoff and Antonia Wachter-Zeh were supported by the German Research Foundation (Deutsche Forschungsgemeinschaft, DFG) under Grant No. WA3907/4-1, the European Research Council (ERC) under the European Union's Horizon 2020 research and innovation program (grant agreement no. 801434), and acknowledge the financial support by the Federal Ministry of Education and Research of Germany in the program of "Souverän. Digital. Vernetzt.". Joint project 6G-life, project identification number: 16KISK002. Patrick Karl acknowledges the financial support by the Federal Ministry of Education and Research of Germany in the program of "Souverän. Digital. Vernetzt.". Joint project 6G-life, project identification number: 16KISK002.

The authors would like to thank Wessel van Woerden and Felicitas Hörmann for pointing out the possible attack on FuLeeca.

References

1. Aguilar-Melchor, C., Gama, N., Howe, J., Hülsing, A., Joseph, D., Yue, D.: The return of the SDitH. Cryptology ePrint Archive (2022)
2. Aragon, N., Blazy, O., Gaborit, P., Hauteville, A., Zémor, G.: Durandal: a rank metric based signature scheme. In: Ishai, Y., Rijmen, V. (eds.) EUROCRYPT 2019. LNCS, vol. 11478, pp. 728–758. Springer, Cham (2019). https://doi.org/10.1007/978-3-030-17659-4_25
3. Aragon, N., Dyseryn, V., Gaborit, P.: Analysis of the security of the PSSI problem and cryptanalysis of the Durandal signature scheme. Cryptology ePrint Archive (2023)
4. Astola, J.: On the asymptotic behaviour of Lee-codes. Discret. Appl. Math. **8**(1), 13–23 (1984)
5. Aumasson, J.P., et al.: SPHINCS$^+$, submission to the NIST post-quantum project, vol. 3 (2020). https://csrc.nist.gov/Projects/post-quantum-cryptography/selected-algorithms-2022
6. Baldi, M., Bianchi, M., Chiaraluce, F., Rosenthal, J., Schipani, D.: Using LDGM codes and sparse syndromes to achieve digital signatures. In: Gaborit, P. (ed.) PQCrypto 2013. LNCS, vol. 7932, pp. 1–15. Springer, Heidelberg (2013). https://doi.org/10.1007/978-3-642-38616-9_1
7. Baldi, M., Bitzer, S., Pavoni, A., Santini, P., Wachter-Zeh, A., Weger, V.: Zero knowledge protocols and signatures from the restricted syndrome decoding problem. Cryptology ePrint Archive (2023)
8. Barenghi, A., Biasse, J.-F., Persichetti, E., Santini, P.: LESS-FM: fine-tuning signatures from the code equivalence problem. In: Cheon, J.H., Tillich, J.-P. (eds.) PQCrypto 2021 2021. LNCS, vol. 12841, pp. 23–43. Springer, Cham (2021). https://doi.org/10.1007/978-3-030-81293-5_2
9. Barg, A.: Complexity issues in coding theory. Technical report TR97-046, Electronic Colloquium on Computational Complexity (ECCC) (1997). https://eccc.weizmann.ac.il/eccc-reports/1997/TR97-046/index.html. ISSN 1433-8092
10. Barg, A.: Some new NP-complete coding problems. Problemy Peredachi Informatsii **30**(3), 23–28 (1994). https://www.mathnet.ru/eng/ppi241
11. Bariffi, J., Bartz, H., Liva, G., Rosenthal, J.: On the properties of error patterns in the constant Lee weight channel. In: International Zurich Seminar on Information and Communication (IZS 2022) Proceedings, pp. 44–48. ETH Zurich (2022)
12. Bariffi, J., Khathuria, K., Weger, V.: Information set decoding for Lee-metric codes using restricted balls. In: Deneuville, J.C. (ed.) CBCrypto 2022. LNCS, vol. 13839, pp. 110–136. Springer, Cham (2022). https://doi.org/10.1007/978-3-031-29689-5_7
13. Becker, A., Joux, A., May, A., Meurer, A.: Decoding random binary linear codes in $2^{n/20}$: how 1+1=0 improves information set decoding. In: Pointcheval, D., Johansson, T. (eds.) EUROCRYPT 2012. LNCS, vol. 7237, pp. 520–536. Springer, Heidelberg (2012). https://doi.org/10.1007/978-3-642-29011-4_31
14. Bellare, M., Rogaway, P.: The exact security of digital signatures-how to sign with RSA and Rabin. In: Maurer, U. (ed.) EUROCRYPT 1996. LNCS, vol. 1070, pp. 399–416. Springer, Heidelberg (1996). https://doi.org/10.1007/3-540-68339-9_34
15. Berlekamp, E.R., McEliece, R.J., van Tilborg, H.C.A.: On the inherent intractability of certain coding problems. IEEE Trans. Inf. Theory **24**(3), 384–386 (1978)
16. Bernstein, D.J.: Grover vs. McEliece. In: Sendrier, N. (ed.) PQCrypto 2010. LNCS, vol. 6061, pp. 73–80. Springer, Heidelberg (2010). https://doi.org/10.1007/978-3-642-12929-2_6

17. Beullens, W.: Sigma protocols for MQ, PKP and SIS, and fishy signature schemes. In: Canteaut, A., Ishai, Y. (eds.) EUROCRYPT 2020. LNCS, vol. 12107, pp. 183–211. Springer, Cham (2020). https://doi.org/10.1007/978-3-030-45727-3_7
18. Bhattacharyya, M., Raina, A.: A quantum algorithm for syndrome decoding of classical error-correcting linear block codes. In: 2022 IEEE/ACM 7th Symposium on Edge Computing (SEC), pp. 456–461 (2022). https://doi.org/10.1109/SEC54971.2022.00069
19. Bidoux, L., Gaborit, P.: Shorter signatures from proofs of knowledge for the SD, MQ, PKP and RSD problems. arXiv preprint arXiv:2204.02915 (2022)
20. Blömer, J., Naewe, S.: Sampling methods for shortest vectors, closest vectors and successive minima. Theoret. Comput. Sci. **410**(18), 1648–1665 (2009)
21. Byrne, E., Horlemann, A.L., Khathuria, K., Weger, V.: Density of free modules over finite chain rings. Linear Algebra Appl. **651**, 1–25 (2022)
22. Chailloux, A., Debris-Alazard, T., Etinski, S.: Classical and Quantum algorithms for generic Syndrome Decoding problems and applications to the Lee metric (2021). https://eprint.iacr.org/2021/552. Report Number: 552
23. Cho, J., No, J.S., Lee, Y., Koo, Z., Kim, Y.S.: Enhanced pqsigRM: code-based digital signature scheme with short signature and fast verification for post-quantum cryptography. Cryptology ePrint Archive (2022)
24. Courtois, N.T., Finiasz, M., Sendrier, N.: How to achieve a McEliece-based digital signature scheme. In: Boyd, C. (ed.) ASIACRYPT 2001. LNCS, vol. 2248, pp. 157–174. Springer, Heidelberg (2001). https://doi.org/10.1007/3-540-45682-1_10
25. Dachman-Soled, D., Ducas, L., Gong, H., Rossi, M.: LWE with side information: attacks and concrete security estimation. In: Micciancio, D., Ristenpart, T. (eds.) CRYPTO 2020. LNCS, vol. 12171, pp. 329–358. Springer, Cham (2020). https://doi.org/10.1007/978-3-030-56880-1_12
26. Debris-Alazard, T., Sendrier, N., Tillich, J.-P.: Wave: a new family of trapdoor one-way preimage sampleable functions based on codes. In: Galbraith, S.D., Moriai, S. (eds.) ASIACRYPT 2019. LNCS, vol. 11921, pp. 21–51. Springer, Cham (2019). https://doi.org/10.1007/978-3-030-34578-5_2
27. Deneuville, J.-C., Gaborit, P.: Cryptanalysis of a code-based one-time signature. Des. Codes Crypt. **88**(9), 1857–1866 (2020). https://doi.org/10.1007/s10623-020-00737-8
28. Ducas, L., et al.: Crystals-dilithium - algorithm specifications and supporting documentation (version 3.1) (2021). https://pq-crystals.org/dilithium/resources.shtml
29. Dumer, I.I.: Two decoding algorithms for linear codes. Problemy Peredachi Informatsii **25**(1), 24–32 (1989)
30. Espitau, T., Tibouchi, M., Wallet, A., Yu, Y.: Shorter hash-and-sign lattice-based signatures. In: Dodis, Y., Shrimpton, T. (eds.) CRYPTO 2022. LNCS, vol. 13508, pp. 245–275. Springer, Cham (2022). https://doi.org/10.1007/978-3-031-15979-4_9
31. Faugere, J.C., Gauthier-Umana, V., Otmani, A., Perret, L., Tillich, J.P.: A distinguisher for high-rate McEliece cryptosystems. IEEE Trans. Inf. Theory **59**(10), 6830–6844 (2013)
32. Feneuil, T.: Building MPCitH-based signatures from MQ, MinRank. Rank SD and PKP. Cryptology ePrint Archive (2022)
33. Feneuil, T., Joux, A., Rivain, M.: Shared permutation for syndrome decoding: new zero-knowledge protocol and code-based signature. Des. Codes Cryptogr. **91**, 1–46 (2022)
34. Feneuil, T., Joux, A., Rivain, M.: Syndrome decoding in the head: Shorter signatures from zero-knowledge proofs. Cryptology ePrint Archive (2022)

35. Fiat, A., Shamir, A.: How to prove yourself: practical solutions to identification and signature problems. In: Odlyzko, A.M. (ed.) CRYPTO 1986. LNCS, vol. 263, pp. 186–194. Springer, Heidelberg (1987). https://doi.org/10.1007/3-540-47721-7_12
36. Fouque, P.A., et al.: FALCON: fast-fourier lattice-based compact signatures over NTRU, specification v1.2 (2020). https://csrc.nist.gov/Projects/post-quantum-cryptography/selected-algorithms-2022
37. Gardy, D., Solé, P.: Saddle point techniques in asymptotic coding theory. In: Cohen, G., Lobstein, A., Zémor, G., Litsyn, S. (eds.) Algebraic Coding 1991. LNCS, vol. 573, pp. 75–81. Springer, Heidelberg (1992). https://doi.org/10.1007/BFb0034343
38. Gentry, C.: Key recovery and message attacks on NTRU-composite. In: Pfitzmann, B. (ed.) EUROCRYPT 2001. LNCS, vol. 2045, pp. 182–194. Springer, Heidelberg (2001). https://doi.org/10.1007/3-540-44987-6_12
39. Gentry, C., Peikert, C., Vaikuntanathan, V.: Trapdoors for hard lattices and new cryptographic constructions. In: Proceedings of the Fortieth Annual ACM Symposium on Theory of Computing, pp. 197–206 (2008)
40. Gligoroski, D., Samardjiska, S., Jacobsen, H., Bezzateev, S.: McEliece in the world of Escher. Cryptology ePrint Archive (2014)
41. Gueron, S., Persichetti, E., Santini, P.: Designing a practical code-based signature scheme from zero-knowledge proofs with trusted setup. Cryptography **6**(1), 5 (2022)
42. Horlemann-Trautmann, A.L., Weger, V.: Information set decoding in the Lee metric with applications to cryptography. Adv. Math. Commun. **15**(4) (2021)
43. Jang, K., Baksi, A., Kim, H., Song, G., Seo, H., Chattopadhyay, A.: Quantum analysis of AES. Cryptology ePrint Archive (2022)
44. Löndahl, C., Johansson, T., Koochak Shooshtari, M., Ahmadian-Attari, M., Aref, M.R.: Squaring attacks on McEliece public-key cryptosystems using quasi-cyclic codes of even dimension. Des. Codes Crypt. **80**, 359–377 (2016)
45. May, A., Meurer, A., Thomae, E.: Decoding random linear codes in $\tilde{\mathcal{O}}(2^{0.054n})$. In: Lee, D.H., Wang, X. (eds.) ASIACRYPT 2011. LNCS, vol. 7073, pp. 107–124. Springer, Heidelberg (2011). https://doi.org/10.1007/978-3-642-25385-0_6
46. Moody, D., Perlner, R.: Vulnerabilities of "McEliece in the world of escher". In: Takagi, T. (ed.) PQCrypto 2016. LNCS, vol. 9606, pp. 104–117. Springer, Cham (2016). https://doi.org/10.1007/978-3-319-29360-8_8
47. Persichetti, E.: Efficient one-time signatures from quasi-cyclic codes: a full treatment. Cryptography **2**(4), 30 (2018)
48. Phesso, A., Tillich, J.-P.: An efficient attack on a code-based signature scheme. In: Takagi, T. (ed.) PQCrypto 2016. LNCS, vol. 9606, pp. 86–103. Springer, Cham (2016). https://doi.org/10.1007/978-3-319-29360-8_7
49. Prange, E.: The use of information sets in decoding cyclic codes. IRE Trans. Inf. Theory **8**(5), 5–9 (1962)
50. Regev, O., Rosen, R.: Lattice problems and norm embeddings. In: Proceedings of the Thirty-Eighth Annual ACM Symposium on Theory of Computing, pp. 447–456 (2006)
51. Santini, P., Baldi, M., Chiaraluce, F.: Cryptanalysis of a one-time code-based digital signature scheme. In: 2019 IEEE International Symposium on Information Theory (ISIT), pp. 2594–2598. IEEE (2019)
52. Schnorr, C.P., Euchner, M.: Lattice basis reduction: improved practical algorithms and solving subset sum problems. Math. Program. **66**, 181–199 (1994)
53. Sendrier, N.: Decoding one out of many. In: Yang, B.-Y. (ed.) PQCrypto 2011. LNCS, vol. 7071, pp. 51–67. Springer, Heidelberg (2011). https://doi.org/10.1007/978-3-642-25405-5_4

54. National Institute of Standards and Technology: SHA-3 Standard: Permutation-Based Hash and Extendable-Output Functions. Technical report (2015). https://doi.org/10.6028/nist.fips.202
55. Stern, J.: A method for finding codewords of small weight. Coding Theory Appl. **388**, 106–113 (1989)
56. Weger, V., Khathuria, K., Horlemann, A.L., Battaglioni, M., Santini, P., Persichetti, E.: On the hardness of the Lee syndrome decoding problem. Adv. Math. Commun. (2022). https://doi.org/10.3934/amc.2022029. https://www.aimsciences.org/en/article/doi/10.3934/amc.2022029

Algebraic Algorithm for the Alternating Trilinear Form Equivalence Problem

Lars Ran(✉), Simona Samardjiska, and Monika Trimoska

Radboud Universiteit, Nijmegen, The Netherlands
{lran,simonas,mtrimoska}@cs.ru.nl

Abstract. The Alternating Trilinear Form Equivalence (ATFE) problem was recently used as a hardness assumption in the design of a digital signature scheme by Tang et al. using the Fiat-Shamir paradigm. It is a hard equivalence problem known to be in the class of equivalence problems that includes, for instance, the Tensor Isomorphism (TI), Quadratic Maps Linear Equivalence (QMLE) and the Matrix Code Equivalence (MCE) problems. Due to the increased cryptographic interest, the understanding of its practical hardness has also increased in the last couple of years. Currently, there are several combinatorial and algebraic algorithms for solving it, the best of which is a graph-theoretic algorithm that also includes an algebraic subroutine.

In this paper, we take a purely algebraic approach to the ATFE problem, but we use a coding theory perspective to model the problem. This modelling was introduced earlier for the MCE problem. Using it, we improve the cost of an algebraic attack against ATFE compared to previously known ones.

Taking into account the algebraic structure of alternating trilinear forms, we show that the obtained system has less variables but also less equations than for MCE and gives rise to structural degree-3 syzygies. Under the assumption that outside of these syzygies the system behaves semi-regularly, we provide a concrete, non-asymptotic complexity estimate of the performance of our algebraic attack. Our results show that the complexity is below the estimated security levels of the signature scheme of Tang et al. and comparable to the currently best graph-theoretic attack by Beullens.

Keywords: trilinear form · matrix codes · algebraic cryptanalysis

1 Introduction

NIST's announcement of reopening the call for post-quantum digital signature proposals specifies the need for shorter signatures whose security is based on problems outside the realm of structured lattices. One family of problems that has recently been brought to focus by the quest for alternative signatures is the family of equivalence problems. The reason behind the rising interest in these problems is that they typically can be used to construct a cryptographic group

A. Esser and P. Santini (Eds.): CBCrypto 2023, LNCS 14311, pp. 84–103, 2023.
https://doi.org/10.1007/978-3-031-46495-9_5

action. Once we have a cryptographic group action, the vectorization problem is used to build a Sigma protocol that, through the Fiat-Shamir transform, can be transformed into a digital signature scheme. For instance, if we take the set comprised of k-tuples of multivariate polynomials together with the group of isomorphisms acting on this set, then we obtain the cryptographic group action underlying the IP signature scheme proposed by Patarin [Pat96]. The original proposition of this scheme is based on the inhomogenous quadratic variant of the isomorphism of polynomials (IP) problem, that is, the case where the polynomials have quadratic, linear and constant terms. This subclass of IP turned out to be easy to solve in practice and hence the IP signature scheme is considered to be broken [FP06]. However, this would not be case for the IP signature scheme instantiated with another subclass of IP, such as the homogenous quadratic variant, also referred to as the Quadratic Maps Linear Equivalence (QMLE) problem.

Recently, as a result of several optimization techniques [DFG19,BKP20, BMPS20,BBPS21] Patarin's construction became attractive again. It was revived through two new signature schemes based on the hardness of two problems closely related to QMLE. A signature scheme based on the hardness of the alternating trilinear form equivalence (ATFE) problem was introduced at Eurocrypt 2022 [TDJ+22], whereas matrix code equivalence (MCE) was used in the more recently proposed construction called MEDS [CNP+22].

As the result of this attention, the understanding of the practical hardness of both ATFE and MCE also significantly improved. An adaptation of Bouillaguet et al. graph-theoretic algorithm for the IP problem [BFV13,Bou11] to the case of ATFE provides an upper bound of $\tilde{\mathcal{O}}(q^{2n/3})$ and this one was used to choose parameters for the scheme from [TDJ+22]. The authors of [TDJ+22] also analyzed the problem purely algebraically, but their model and assumptions on the obtained algebraic system gave worse estimates of $\mathcal{O}(2^{6\omega n\log_2(n)})$. They further provided a basic collision based approach similar in nature to the one in [BFV13] but looking at low rank codewords as in Leon's algorithm for the Hamming metric [Leo82,Beu20]. This basic attack was subsequently improved by Beullens [Beu22] to $\tilde{\mathcal{O}}(q^{\max{(n-5)/2, n-7}})$ for odd n and $\tilde{\mathcal{O}}(q^{\max{(n-4)/2, n-4}})$ for even n. For some special cases of weak keys, even better results were presented leading to practical polynomial time attacks. If such weak keys are avoided, the attack performs better in the odd n case.

Similar approaches were taken into account for the MCE problem which was analyzed in [RST22,CNP+22]. In [RST22], Bouillaguet et al.'s algorithm was transformed into an algorithm of complexity $\tilde{\mathcal{O}}(q^{4n/3})$. Using a different property for building the graph, the authors proposed an improvement resulting in a complexity of $\tilde{\mathcal{O}}(q^n)$. Currently the best algorithms against MCE were developed in [CNP+22] and they take nontrivial approaches in adapting Leon's algorithm to the rank metric and modeling the problem algebraically but from a coding theory viewpoint. The focus of this work is related to this improved algebraic modeling.

Our Contribution. In this work, we take advantage of the relation between the two equivalence problems – ATFE and MCE to improve the cryptanalysis of ATFE. Namely, an alternating trilinear form can easily be represented as a matrix code. A reduction from ATFE to MCE directly follows from the reduction results in [TDJ+22] and [RST22]. Theorem 2 in [TDJ+22] states that ATFE is tensor isomorphism complete and thus equivalent to QMLE and Theorem 11 in [RST22] shows a reduction from QMLE (the bilinear case) to MCE. Specifically, an MCE instance with a pair of matrix codes derived from a positive ATFE instance, has a solution of the form $(\mathbf{A}^\top, \mathbf{A})$, where the matrix $\mathbf{A}$ is a solution to the original ATFE instance.

Viewing ATFE as a problem on matrix codes enables us to model the problem using coding theory techniques. In particular, we model ATFE algebraically in a nontrivial way using the approach from [CNP+22] for MCE. This model improves the cost of an algebraic attack compared to previously known models as for example described in [TDJ+22].

Taking into account the algebraic structure of alternating trilinear forms, we show that the obtained system has less variables but also less equations than for MCE. In particular we can model ATFE as a system of $n(\binom{n}{2} - n)$ equations in n^2 variables.

For our complexity analysis, we first show the existence of $\frac{(n+1)(n-1)(n-3)}{3}$ structural degree-3 syzygies in such systems. Then, under the assumption that outside of these syzygies the system behaves semi-regularly, we show that the complexity is below the estimated security levels of the signature scheme in [TDJ+22] for all proposed parameters, and comparable to the currently best graph-theoretic attack of Beullens [Beu22] against ATFE. Furthermore, we provide concrete, non-asymptotic, security estimates for other, higher parameter sets. Our results for the parameters proposed in [TDJ+22] are given in Table 1.

Table 1. Comparison of the concrete complexities (in log_2 scale) of different algorithms for solving ATFE.

n	q	Tang et al. [TDJ+22]	Beullens [Beu22]	Our work
9	524287	133	38	90
10	131071	133	122	95
11	65521	138	85	101

Organization. The paper is organized as follows. Section 2 introduces the necessary preliminaries and Sect. 3 reviews state-of-the-art algorithms for solving ATFE. In Sect. 4, we show how a positive ATFE instance is transformed into a positive MCE instance and we explore the structure of the matrix codes obtained from this transformation. In Sect. 5 we show algebraic modellings for the ATFE problem and in Sect. 6 we give a complexity analysis for solving the systems from our proposed variant. Finally, in Sect. 7 we show our experimental results.

2 Preliminaries

Let $\mathbb{F}_q$ be the finite field of q elements. $\mathrm{GL}_n(q)$ and $\mathrm{AGL}_n(q)$ denote respectively the general linear group and the general affine group of degree n over $\mathbb{F}_q$. We use bold letters to denote vectors $\mathbf{a}, \mathbf{c}, \mathbf{x}, \ldots$, and matrices $\mathbf{A}, \mathbf{B}, \ldots$. The entries of a vector $\mathbf{a}$ are denoted by a_i, and we write $\mathbf{a} = (a_1, \ldots, a_n)$ for a (column) vector of dimension n over some field. Similarly, the entries of a matrix $\mathbf{A}$ are denoted by a_{ij}. We denote by $\mathbf{e}_1, \ldots, \mathbf{e}_n$ the vectors of the canonical basis of $\mathbb{F}_q^n$. If $\mathbf{b}_1, \ldots, \mathbf{b}_n$ is a basis for a vector space, we denote by $\mathbf{b}_1^*, \ldots, \mathbf{b}_n^*$ the corresponding dual basis. We denote by $\mathcal{S}_n$ the symmetric group of degree n. Finally, we denote the set of all $m \times n$ matrices over $\mathbb{F}_q$ by $\mathcal{M}_{m,n}(\mathbb{F}_q)$.

Cryptographic Group Actions

Definition 1. Let X be a set and $(G, \cdot)$ be a group. A group action is a mapping

$$\begin{aligned} \star : G \times X &\rightarrow X \\ (g, x) &\mapsto g \star x \end{aligned}$$

such that the following conditions hold for all $x \in X$:

- $e \star x = x$, where e is the identity element of G.
- $g_2 \star (g_1 \star x) = (g_2 \cdot g_1) \star x$, for all $g_1, g_2 \in G$.

A *cryptographic* group action commonly refers to a group action that has some additional properties that are useful for cryptographic applications. To begin with, there are some desirable properties of computational nature. Namely, the following procedures should be efficient:

- *Evaluation*: given x and g, compute $g \star x$.
- *Sampling*: sample uniformly at random from G.
- *Membership testing*: verify that $x \in X$.

The crucial property that distinguishes cryptographic group actions is that the corresponding *vectorization problem* should be hard:

Problem 1. GroupActionVectorization(X, x_1, x_2):
Input: The pair $x_1, x_2 \in X$.
Question: Find – if any – $g \in G$ such that $g \star x_1 = x_2$.

Early constructions using this paradigm are based on the action of finite groups of prime order, for which the vectorization problem is the discrete logarithm problem. Notable isogeny-based constructions can be found, for instance, in the work of Couveignes in [Cou06] and later by Rostovtsev and Stolbunov [RS06]. Recently, a general framework based on group actions was explored in more detail by [ADFMP20], allowing for the design of several primitives.

The Alternating Trilinear Form Equivalence Problem. A k-linear form is a function $\phi : \mathbb{F}_q^n \times \cdots \times \mathbb{F}_q^n \to \mathbb{F}_q$ that is linear in each argument: if we fix $k-1$ arguments, it is linear in the remaining argument. A k-linear form is called

- *symmetric*: if $\phi(\mathbf{x}_1, \ldots, \mathbf{x}_k) = \phi(\mathbf{x}_{\pi(1)}, \ldots, \mathbf{x}_{\pi(k)})$ for any permutation $\pi \in \mathcal{S}_k$;
- *skew-symmetric*: if $\phi(\mathbf{x}_1, \ldots, \mathbf{x}_k) = \phi(\mathbf{x}_{\tau(1)}, \ldots, \mathbf{x}_{\tau(k)})$ for any transposition $\tau \in \mathcal{S}_k$;
- *alternating* if $\phi(\mathbf{x}_1, \ldots, \mathbf{x}_k) = 0$ whenever $\mathbf{x}_i = \mathbf{x}_j$ for some $i \neq j$.

Every alternating form is skew-symmetric, and if $q \geq 3$, every skew-symmetric form is alternating. In the following, we will focus on the $k = 2$ and $k = 3$ cases: bilinear and trilinear forms.

An alternating trilinear form can be represented as $\sum_{1 \leqslant i < j < s \leqslant n} c_{ijs}(\mathbf{e}_i^* \wedge \mathbf{e}_j^* \wedge \mathbf{e}_s^*)$, where $c_{ijs} \in \mathbb{F}_q$, $\mathbf{e}_i$ is the ith canonical basis vector, $\mathbf{e}_i^*$ is the linear form sending $u = (u_1, ..., u_n) \in \mathbb{F}_q^n$ to u_i and $\wedge$ denotes the wedge product. Hence, $\mathbf{e}_i^* \wedge \mathbf{e}_j^* \wedge \mathbf{e}_s^*$ is an alternating form sending $(\mathbf{x}, \mathbf{y}, \mathbf{z})$ to the determinant $\begin{vmatrix} x_i & y_i & z_i \\ x_j & y_j & z_j \\ x_s & y_s & z_s \end{vmatrix}$. From this representation it is clear that an alternating trilinear form can be stored using $\binom{n}{3}$ entries: one for each coefficient c_{ijs}.

The alternating trilinear form equivalence problem is formally defined as follows:

Problem 2. ATFE(n, ϕ, ψ):
Input: Two alternating trilinear forms ϕ, ψ.
Question: Find – if any – $\mathbf{A} \in \mathrm{GL}_n(q)$ such that $\psi(\mathbf{x}, \mathbf{y}, \mathbf{z}) = \phi(\mathbf{A}\mathbf{x}, \mathbf{A}\mathbf{y}, \mathbf{A}\mathbf{z})$.

The ATFE-based group action is defined by the action of the general linear group $\mathrm{GL}_n(q)$ on the set of all alternating trilinear forms defined over $\mathbb{F}_q^n$. The vectorization problem is the ATFE problem defined above. Since ATFE is a hard problem, we obtain a cryptographic group action.

Array Representation of Bilinear and Trilinear Forms. It is common to represent a bilinear form as $\mathbf{x}^\top \mathbf{M} \mathbf{y}$, where $\mathbf{M}$ is a matrix where the (i, j) entry holds the coefficient of the term $x_i y_j$. Similarly, trilinear forms can be represented with a 3-way array where the (i, j, s) entry holds the coefficient of $x_i y_j z_s$. In this representation, we implicitly choose $\mathbf{e}_1, \ldots, \mathbf{e}_n$ as a basis for $\mathbb{F}_q^n$. Alternating bilinear and trilinear forms can be represented in such a way, although it is not the most efficient representation. The array representation of an alternating bilinear form is a skew-symmetric matrix with zeros on the main diagonal. The array representation of a trilinear form has even more redundancy. Notice from the 'determinant representation' above that for all permutations of the index triple (i, j, s), the terms $x_i y_j z_s$ have the same coefficient, up to sign. Specifically, if we denote by M_{ijs} the (i, j, s) entry of the 3-way array, then $M_{ijs} = -M_{isj} = M_{sij} = -M_{jis} = M_{jsi} = -M_{sji}$. This is the key property that makes all of the terms cancel out (and hence the form evaluate to zero) whenever two arguments are the same.

The Matrix Code Equivalence Problem. A *matrix code* is a subspace $\mathcal{C}$ of $m \times n$ matrices over $\mathbb{F}_q$ endowed with the rank metric defined as $d(\mathbf{A}, \mathbf{B}) = \text{Rank}(\mathbf{A} - \mathbf{B})$. We denote by k the dimension of $\mathcal{C}$ as a subspace of $\mathbb{F}_q^{m \times n}$ and its basis by $(\mathbf{C}^{(1)}, \ldots, \mathbf{C}^{(k)})$ where $\mathbf{C}^{(i)} \in \mathbb{F}_q^{m \times n}$ are linearly independent.

The matrix code equivalence problem is formally defined as follows:

Problem 3. $\mathsf{MCE}(k, n, m, \mathcal{C}, \mathcal{D})$:
Input: Two k-dimensional matrix codes $\mathcal{C}, \mathcal{D} \subset \mathcal{M}_{m,n}(\mathbb{F}_q)$.
Question: Find – if any – $\mathbf{A} \in \text{GL}_m(q), \mathbf{B} \in \text{GL}_n(q)$ such that for all $\mathbf{C} \in \mathcal{C}$, it holds that $\mathbf{ACB} \in \mathcal{D}$.

Algebraically, the MCE problem corresponds to the problem of finding the unknown entries of matrices $\mathbf{A}, \mathbf{B}, \mathbf{T}$ such that

$$\mathbf{D}^{(i)} = \sum_{1 \leqslant j \leqslant n} t_{ji} \mathbf{AC}^{(j)} \mathbf{B}, \quad \forall i, 1 \leqslant i \leqslant n$$

is satisfied. The matrix $\mathbf{T} \in \text{GL}_k(q)$ corresponds to a change of basis of $\mathbf{A}\mathcal{C}\mathbf{B}$.

The MCE problem also gives rise to a group action: the group $\text{GL}_m(q) \times \text{GL}_n(q)$ acts on the set formed by the k-dimensional matrix codes of size $m \times n$ over the base field $\mathbb{F}_q$. The vectorization problem is MCE, and since this is a hard problem, we obtain a cryptographic group action.

Exterior Powers and Extending Trilinear Forms. For combinatorial analysis it can be useful to work with linear maps instead of trilinear maps. To this end we introduce, for every k, the exterior powers of a vector space. These are vector spaces generated by wedge products:

$$\bigwedge^k \mathbb{F}_q^n := \{\sum_i (\mathbf{x}_1)_i \wedge \ldots \wedge (\mathbf{x}_k)_i \mid (\mathbf{x}_j)_i \in \mathbb{F}_q^n\}.$$

These vector spaces have dimension $\binom{n}{k}$. Furthermore, linear transformations $\mathbf{A} : \mathbb{F}_q^n \to \mathbb{F}_q^n$ also act on $\bigwedge^k \mathbb{F}_q^n$ by

$$\mathbf{A}(\mathbf{x}_1 \wedge \ldots \wedge \mathbf{x}_k) = \mathbf{A}\mathbf{x}_1 \wedge \ldots \wedge \mathbf{A}\mathbf{x}_k.$$

Now each alternating k-linear form $\phi : \mathbb{F}_q^n \times \ldots \times \mathbb{F}_q^n \to \mathbb{F}_q$ can be extended to a linear form $\hat{\phi} : \bigwedge^k \mathbb{F}_q^n \to \mathbb{F}_q$ where the map is given by:

$$\hat{\phi}\left(\sum_i (\mathbf{x}_1)_i \wedge \ldots \wedge (\mathbf{x}_k)_i\right) = \sum_i \phi\left((\mathbf{x}_1)_i, \ldots, (\mathbf{x}_k)_i\right).$$

This extension is unique and is in fact a natural bijection between k-linear forms and linear forms on the kth exterior power. Therefore we will abuse notation and write ϕ for both maps. The number of arguments will indicate what is meant.

This can also be used to partly linearize a k-linear form in the first l arguments. In this case, an alternating k-linear form $\phi : \mathbb{F}_q^n \times \ldots \times \mathbb{F}_q^n \to \mathbb{F}_q$ can be extended to a $(k-l+1)$-linear form

$$\hat{\phi} : \bigwedge^l \mathbb{F}_q^n \times \overbrace{\mathbb{F}_q^n \times \ldots \times \mathbb{F}_q^n}^{k-l \text{ times}} \to \mathbb{F}_q$$
$$(\mathbf{x}_1 \wedge \ldots \wedge \mathbf{x}_l, \mathbf{x}_{l+1}, \ldots \mathbf{x}_k) \mapsto \phi(\mathbf{x}_1, \ldots \mathbf{x}_k).$$

This extension is again unique. Note that this extension has arguments from different spaces so it is not alternating any more. We will again denote both forms by ϕ, the number and type of arguments should indicate what is meant. For our use case, $k = 3$, this implies the following equations:

$$\phi(\mathbf{x}, \mathbf{y}, \mathbf{z}) = \phi(\mathbf{x} \wedge \mathbf{y}, \mathbf{z}) = \phi(\mathbf{x} \wedge \mathbf{y} \wedge \mathbf{z}).$$

For a more thorough treatment on exterior powers, alternating forms and multilinear algebra in general we refer the reader to [Gre12].

3 Previous Algorithms for Solving ATFE

The state-of-the-art algorithms against ATFE build upon relatively old algorithms against the Isomorphism of polynomials (IP) [Per05, BFFP11, BFV13]. We present the two most relevant below.

3.1 Graph-Theoretic Algorithm of Bouillaguet et al. [BFV13]

More than 10 years ago, Bouillaguet et al. [BFV13] proposed a birthday-based graph-theoretic algorithm for solving the Quadratic Maps Linear Equivalence (QMLE) problem. It is now known that the ATFE problem is polynomial-time equivalent to the homogeneous version of QMLE [GQ21] implying that this algorithm can be adapted for ATFE.

Specifically, two isomorphic alternating trilinear forms ϕ and ψ over $\mathbb{F}_q^n$ can be seen as two equivalent homogeneous quadratic maps $\mathcal{F}$ and $\mathcal{P}$ of n multivariate polynomials in n variables over $\mathbb{F}_q$. Furthermore, these quadratic maps are alternating and bilinear, so they have a skew-symmetric matrix representation. The main observation of the algorithm is that once a pair of vectors $\mathbf{u}, \mathbf{v} \in \mathbb{F}_q^n$ is known such that $\mathbf{u} = \mathbf{A}\mathbf{v}$, this information is enough to find the isomorphism with low complexity[1]. Hence, the goal of the algorithm is to find this collision of points, and different invariants under isomorphism can be used to achieve this.

[1] In [BFV13] it was conjectured that this complexity is $\mathcal{O}(n^9)$ i.e. polynomial. Later in [Bou11, RST22] this was reevaluated and shown that the conclusion was made based on some false assumptions. Nevertheless, even though there is no proof of the polynomial behavior of this step, in practice it does finish in an expected polynomial time.

For the case of ATFE, a useful invariant is the rank of the corresponding bilinear form $\phi_{\mathbf{v}}(\mathbf{w}, \mathbf{z}) = \phi(\mathbf{v}, \mathbf{w}, \mathbf{z})$ which is preserved under the isomorphism defined by $\mathbf{A}$. The algorithm now proceeds as a standard collision-search algorithm in two steps: First, create lists L_ϕ and L_ψ of size $\mathcal{O}(q^{n/3})$ elements in $\mathbb{F}_q^n$ of the same rank. Then, find a collision between these lists by calling the efficient algorithm described above. The total complexity amounts to $\tilde{\mathcal{O}}(q^{2n/3})$ where we neglect the estimated $\mathcal{O}(n^9)$ cost of finding the isomorphism once one collision is known.

3.2 Graph-Theoretic Algorithm of Beullens [Beu22]

Beullens [Beu22] improves generically upon the previous approach by further using clever graph-walking techniques. The basic idea is to populate the lists faster by exploiting the structure of a particular invariant graph for alternating trilinear forms. This graph had been studied before and was used for complete classification of trilinear forms of dimensions $n = 8, 9$. Namely, the structure of the graph allows to find points of the same or lower rank in the neighborhood of an identified point of a specified rank in polynomial time. Thus, one can first find using brute force a point of higher rank (which is easier than finding one of lower rank), and then by exploring the neighborhood can find points of lower rank faster. In total, this costs $\tilde{\mathcal{O}}(q^{(n-5)/2})$ for odd n and $\tilde{\mathcal{O}}(q^{(n-4)/2})$ for even n. The second part of the algorithm is as previous and consists of matching each pair in the lists and checking whether it leads to the unknown isomorphism. This part has a complexity of $\tilde{\mathcal{O}}(q^{n-7})$ for odd n and $\tilde{\mathcal{O}}(q^{n-4})$ for even n, and for larger n, it becomes the dominating part of the algorithm.

4 A Coding Theory Perspective of ATFE

A trilinear form can be seen as a matrix code and the other way around.

For an informal argument for the equivalence between these two objects, we refer to their algorithmic representation. A matrix code is usually represented by an array of the matrices forming its basis. This is a 3-way array, no different than a 3-way array representing a trilinear form as described in Sect. 2. It is then evident that we can obtain a matrix code from an (alternating) trilinear form simply by choosing a basis for the code.

Indeed, let $\phi^{(i)}(\mathbf{x}, \mathbf{y}) = \phi(\mathbf{x}, \mathbf{y}, \mathbf{e}_i)$ be the bilinear form obtained by fixing the third argument of a trilinear form ϕ to $\mathbf{e}_i$, where $\mathbf{e}_i$ denotes the ith vector of the canonical basis $(\mathbf{e}_1, \ldots, \mathbf{e}_n)$. With respect to this basis, a vector $\mathbf{a} = \sum \alpha_i \mathbf{e}_i$ can be written as $\mathbf{a} = (\alpha_1, \ldots, \alpha_n)$. If ϕ is alternating, then $\phi^{(i)}$ is also alternating and it holds that $\phi^{(i)}(-, \mathbf{e}_i) = \phi^{(i)}(\mathbf{e}_i, -) = 0$. Let $\mathbf{C}^{(i)}$ be the matrix representation of $\phi^{(i)}$. Then, $(\mathbf{C}^{(1)}, \ldots, \mathbf{C}^{(n)})$ is a basis of an n-dimensional matrix code. The only piece left is to show the relation between the solutions of two such related instances. Specifically, we show the following.

Lemma 1. *Finding a solution of the form $(\mathbf{A}^\top, \mathbf{A})$ to an* MCE *instance derived from an* ATFE *instance is equivalent to finding a solution* $\mathbf{A}$ *to the original* ATFE *instance.*

Proof. Let (n, ϕ, ψ) be an instance of ATFE and let $\mathcal{C}$ and $\mathcal{D}$ be matrix codes obtained by applying the above transformation to ϕ and ψ respectively. If (n, ϕ, ψ) is a positive instance of ATFE, then there exists $\mathbf{A} \in \mathrm{GL}_n(q)$ such that $\psi^{(i)}(\mathbf{x}, \mathbf{y}) = \psi(\mathbf{x}, \mathbf{y}, \mathbf{e}_i) = \phi(\mathbf{Ax}, \mathbf{Ay}, \mathbf{Ae}_i)$ for all $i \in \{1, \ldots, n\}$. Since $\mathbf{Ae}_i = (a_{1i}, \ldots, a_{ni})$, we have that $\psi^{(i)}(\mathbf{x}, \mathbf{y}) = \phi(\mathbf{Ax}, \mathbf{Ay}, a_{1i}\mathbf{e}_1 + \cdots + a_{ni}\mathbf{e}_n)$. By linearity, we infer that $\psi^{(i)}(\mathbf{x}, \mathbf{y}) = \sum_{1 \leqslant j \leqslant n} a_{ji}\phi(\mathbf{Ax}, \mathbf{Ay}, \mathbf{e}_j) = \sum_{1 \leqslant j \leqslant n} a_{ji}\phi^{(j)}(\mathbf{Ax}, \mathbf{Ay})$. This can be rewritten in matrix form as $\mathbf{x}^\top \mathbf{D}^{(i)} \mathbf{y} = \sum_{1 \leqslant j \leqslant n} a_{ji}(\mathbf{Ax})^\top \mathbf{C}^{(j)}(\mathbf{Ay})$, $\forall i, 1 \leqslant i \leqslant n$. Since this holds for any $(\mathbf{x}, \mathbf{y})$, we have that

$$\mathbf{D}^{(i)} = \sum_{1 \leqslant j \leqslant n} a_{ji} \mathbf{A}^\top \mathbf{C}^{(j)} \mathbf{A}, \quad \forall i, 1 \leqslant i \leqslant n. \tag{1}$$

Taking $(\mathbf{C}^{(1)}, \ldots, \mathbf{C}^{(n)})$ as a basis of a matrix code $\mathcal{C}$ and $(\mathbf{D}^{(1)}, \ldots, \mathbf{D}^{(n)})$ as a basis of a matrix code $\mathcal{D}$, from Eq. (1) we infer that

- The codes $\mathcal{C}$ and $\mathcal{D}$ are equivalent up to a change of basis represented by the matrix $\mathbf{A}$.
- $(\mathbf{A}^\top, \mathbf{A})$ is a solution to the MCE instance $(n, n, n, \mathcal{C}, \mathcal{D})$. □

Example 1. Let

$$\begin{aligned}\phi(\mathbf{x}, \mathbf{y}, \mathbf{z}) = {} & x_2y_3z_1 + 3x_2y_4z_1 + 6x_3y_2z_1 + 6x_3y_4z_1 + 4x_4y_2z_1 + x_3y_4z_1 \\ & + 6x_1y_3z_2 + 4x_1y_4z_2 + x_3y_1z_2 + 6x_3y_4z_2 + 3x_4y_1z_2 + x_4y_3z_2 + x_1y_2z_3 \\ & + x_1y_4z_3 + 6x_2y_1z_3 + x_2y_4z_3 + 6x_4y_1z_3 + 6x_4y_2z_3 + 3x_1y_2z_4 + 6x_1y_3z_4 \\ & + 4x_2y_1z_4 + 6x_2y_3z_4 + x_3y_1z_4 + x_3y_2z_4\end{aligned}$$

and

$$\begin{aligned}\psi(\mathbf{x}, \mathbf{y}, \mathbf{z}) = {} & 6x_2y_3z_1 + 6x_2y_4z_1 + x_3y_2z_1 + x_4y_2z_1 + x_1y_3z_2 + x_1y_4z_2 \\ & + 6x_3y_1z_2 + 6x_3y_4z_2 + 6x_4y_1z_2 + x_4y_3z_2 + 6x_1y_2z_3 + x_2y_1z_3 + x_2y_4z_3 \\ & + 6x_4y_2z_3 + 6x_1y_2z_4 + x_2y_1z_4 + 6x_2y_3z_4 + x_3y_2z_4\end{aligned}$$

be two equivalent alternating trilinear forms over $\mathbb{F}_7$. The terms that are redundant in a compact representation are written in green. An isomorphism between these two forms is, for instance,

$$\mathbf{A} = \begin{pmatrix} 6&4&5&1 \\ 2&0&2&0 \\ 1&2&6&2 \\ 5&6&6&1 \end{pmatrix}.$$

The corresponding codes are

$$\mathcal{C} = \left(\begin{pmatrix} 0&0&0&0 \\ 0&0&1&3 \\ 0&6&0&6 \\ 0&4&1&0 \end{pmatrix}, \begin{pmatrix} 0&0&6&4 \\ 0&0&0&0 \\ 1&0&0&6 \\ 3&0&1&0 \end{pmatrix}, \begin{pmatrix} 0&1&0&1 \\ 6&0&0&1 \\ 0&0&0&0 \\ 6&6&0&0 \end{pmatrix}, \begin{pmatrix} 0&3&6&0 \\ 4&0&6&0 \\ 1&1&0&0 \\ 0&0&0&0 \end{pmatrix} \right)$$

and

$$\mathcal{D} = \left(\begin{pmatrix} 0&0&0&0 \\ 0&0&6&6 \\ 0&1&0&0 \\ 0&1&0&0 \end{pmatrix}, \begin{pmatrix} 0&0&1&1 \\ 0&0&0&0 \\ 6&0&0&6 \\ 6&0&1&0 \end{pmatrix}, \begin{pmatrix} 0&6&0&0 \\ 1&0&0&1 \\ 0&0&0&0 \\ 0&6&0&0 \end{pmatrix}, \begin{pmatrix} 0&6&0&0 \\ 1&0&6&0 \\ 0&1&0&0 \\ 0&0&0&0 \end{pmatrix} \right).$$

We can check that $(\mathbf{A}^\top, \mathbf{A})$ is an isometry from $\mathcal{C}$ to $\mathcal{D}$. Note that for such small parameters ($n = 4$), there are probably many isometries from $\mathcal{C}$ to $\mathcal{D}$.

The codes $\mathcal{C}$ and $\mathcal{D}$ have several properties intrinsic to their derivation from alternating trilinear forms. For simplicity, we discuss all of them assuming the choice of basis specified in the beginning of this section. All of the matrices forming the basis of $\mathcal{C}$ are skew-symmetric with zeros on the main diagonal, hence they are all of even rank. More generally, we have the following relations between their entries: $C_{ij}^{(s)} = -C_{is}^{(j)} = C_{si}^{(j)} = -C_{ji}^{(s)} = C_{js}^{(i)} = -C_{sj}^{(i)}$. The same holds for the basis of $\mathcal{D}$. The ith column and the ith row is zero in the ith matrix of the basis, that is, the matrix corresponding to the bilinear form $\phi^{(i)}(\mathbf{x}, \mathbf{y})$ (resp. $\psi^{(i)}(\mathbf{x}, \mathbf{y})$ for $\mathcal{D}$). These zero column and row vectors, as well as the zeros on the diagonal, result from the property that in an alternating trilinear form, the coefficient of a term $x_i y_j z_s$ is zero if any two of the three indices (i, j, s) are the same. Finally, positive MCE instances derived from positive ATFE instances have a specific solution. Instead of a pair of unrelated matrices, we have a solution $(\mathbf{A}, \mathbf{B})$ such that $\mathbf{A} = \mathbf{B}^\top$. Hence ATFE can be reduced to a subclass of MCE.

5 Algebraic Algorithms for Solving ATFE

In view of the connection of ATFE to MCE we continue to use the matrix code representation introduced in the previous section.

5.1 Direct Modelling

A straightforward way to model this problem algebraically is to describe Eq. (1) as a system of $n \cdot \binom{n}{2}$ equations in n^2 variables, corresponding to the coefficients of $\mathbf{A}$. The resulting system is of degree three. Alternatively, we can move one linear transformation to the other side of the equality and obtain

$$\sum_{1 \leqslant j \leqslant n} \tilde{a}_{ji} \mathbf{D}^{(i)} = \mathbf{A}^\top \mathbf{C}^{(j)} \mathbf{A}, \quad \forall i, 1 \leqslant i \leqslant n, \tag{2}$$

where $\tilde{a}_{ji}$ is the (j, i) entry of $\mathbf{A}^{-1}$. When we rewrite the system like this, the number of equations does not change and we double the number of variables, but we obtain an inhomogenous quadratic system instead of a cubic one. Specifically, the system is quadratic in the $\mathbf{A}$-variables and linear in the $\mathbf{A}^{-1}$-variables. We add to this the constraint $\mathbf{A}\mathbf{A}^{-1} = \mathbf{I}$, which yields n^2 equations that are bilinear in $\mathbf{A}$-variables and $\mathbf{A}^{-1}$-variables. We will refer to this approach as the *direct* modelling. The direct modelling dates back to the work in [FP06] for solving the QMLE problem, with further analysis in [BFV13,Bou11]. Recently, it was analysed as a modelling for MCE in [CNP+22], before a more advanced approach was introduced. The work in this paper shows that the improved modelling introduced in [CNP+22] is even more relevant in the ATFE case. We describe this approach in the following subsection.

A similar modelling was used in [TDJ+22] for the analysis of an algebraic attack on ATFE. In fact, with the algebraic modelling in [TDJ+22] we obtain a subset of the equations in the system arising from Eq. (2). Due to the compact representation of ATFE, the number of equations is $\binom{n}{3} + n^2$, which is less than the $n \cdot \binom{n}{2} + n^2$ equations that we obtain from the corresponding matrix representation. The complexity of this approach is analysed under the assumption that the polynomials in the system form a semi-regular sequence. Using the analysis techniques from [Bar04,BFSY05], the degree of regularity is estimated to be $3n$ asymptotically, and the complexity is upper-bounded by $\mathcal{O}(N^{3n\omega})$, where $N = 2n^2$ is the number of variables and ω is the linear algebra constant.

In [BDN+23], the direct modelling is improved by adding the equations arising from

$$\sum_{1 \leqslant j \leqslant n} a_{ji}\mathbf{C}^{(i)} = (\mathbf{A}^{-1})^{\top}\mathbf{D}^{(j)}\mathbf{A}^{-1}, \quad \forall i, 1 \leqslant i \leqslant n,$$

and also $\mathbf{A}^{-1}\mathbf{A} = \mathbf{I}$. This is called the *quadratic with inverse* modelling and results in a system of $2n(\binom{n}{2} + n)$ equations in $2n^2$ variables. In [BDN+23], it is used as reference for calculating the complexity of an algebraic attack on the ATFE problem.

5.2 Improved Matrix-Code Modelling

The improved modelling uses ideas from coding theory and its greatest advantage is that all variables that occur linearly in the direct modelling are not present in the improved system. In this description of the modelling, we will focus on MCE instances derived from ATFE instances. For these instances, we obtain a polynomial system in n^2 variables, which is a significant improvement over the system with $2n^2$ variables obtained from the direct modelling.

Let $\mathbf{G}$ and $\mathbf{G}'$ be the $n \times n^2$ generator matrices of $\mathcal{C}$ and $\mathcal{D}$ respectively. These generator matrices are obtained by *flattening* the matrix code, in the following manner. For a matrix $\mathbf{C} \in \mathcal{M}_{n,n}(\mathbb{F}_q)$, let vec be a mapping that sends a matrix $\mathbf{C}$ to the vector $\mathsf{vec}(\mathbf{C}) \in \mathbb{F}_q^{n^2}$ obtained by:

$$\mathsf{vec}: \ \mathbf{C} = \begin{pmatrix} a_{1,1} & \dots & a_{1,n} \\ \vdots & \ddots & \vdots \\ a_{n,1} & \dots & a_{n,n} \end{pmatrix} \mapsto \mathsf{vec}(\mathbf{C}) = (a_{1,1}, \dots, a_{1,n}, \dots, a_{n,1}, \dots, a_{n,n}).$$

Then $\mathbf{G}$ is constructed as follows

$$\mathbf{G} := \begin{pmatrix} \mathsf{vec}(\mathbf{C}_1) \\ \vdots \\ \mathsf{vec}(\mathbf{C}_n) \end{pmatrix}.$$

The representation using generator matrices constructed as above allows us to view a matrix code as an $\mathbb{F}_q$-subspace of $\mathbb{F}_q^{n^2}$. We can now describe the improved modelling in three steps:

- Compute $\mathbf{G}'^{\perp}$, that is, the generator matrix of the dual code of $\mathcal{D}$. This is an $(n^2 - n) \times n^2$ matrix containing only constant values, and it can be computed directly from $\mathbf{G}'$.
- Compute $\tilde{\mathbf{G}}$, that is, a generator matrix of $\mathcal{D}$ represented as $\mathbf{A}^\top \mathcal{C} \mathbf{A}$ for $\mathbf{A}$ with unknown coefficients. This is an $n \times n^2$ matrix whose entries are quadratic equations in the $\mathbf{A}$-variables. It can be obtained either by computing matrices $\mathbf{A}^\top \mathbf{C}_i \mathbf{A}$ and flattening them to obtain the rows of $\tilde{\mathbf{G}}$, or by computing $\tilde{\mathbf{G}} = \mathbf{G}(\mathbf{A} \otimes \mathbf{A})$.
- Construct the system

$$\mathbf{G}'^{\perp} \cdot \tilde{\mathbf{G}}^\top = \mathbf{0}_{(n^2-n)\times n}. \tag{3}$$

Note that the system obtained from Eq. (3) has $n(n^2 - n)$ equations, but only $n(\binom{n}{2} - n)$ of them are linearly independent because of the specific structure of matrix codes obtained from alternating trilinear forms. Recall from Sect. 4 that we have the following relations between the entries of the matrices from the basis: $C_{ij}^{(s)} = -C_{is}^{(j)} = C_{si}^{(j)} = -C_{ji}^{(s)} = C_{js}^{(i)} = -C_{sj}^{(i)}$. This shows that any generator matrix $\mathbf{G}$ of a matrix code derived from an alternating trilinear form has $\binom{n}{2}$ linearly independent columns. For an alternative view of this modelling that is in the spirit of the minors modellings of MinRank [FdVP08,BBC+20], we refer the reader to [CNP+22].

5.3 Removing Invalid Solutions

One drawback of the improved modelling is that it does not contain the constraint that the solution $\mathbf{A}$ has to be an invertible matrix. As a consequence, the polynomial system can have solutions that do not correspond to solutions to the ATFE instance, and this effect can significantly slow down the resolution of the system. Note that the direct modelling does not have this problem because there are equations describing $\mathbf{A}\mathbf{A}^{-1} = \mathbf{I}$.

As an example for invalid solutions we show that all rank-1 matrices $\mathbf{A}$ are a solution to the improved modelling as is. Let $\mathbf{A} = \mathbf{a}\mathbf{b}^\top$, then $\mathbf{A}^\top \mathbf{C}_i \mathbf{A} = \mathbf{b}\mathbf{a}^\top \mathbf{C}_i \mathbf{a}\mathbf{b}^\top$. But we know that $\mathbf{C}_i$ is skew-symmetric, hence $\mathbf{a}^\top \mathbf{C}_i \mathbf{a} = 0$. After flattening, $\tilde{G} = \mathbf{0}$ and our system is trivially satisfied.

In the following, we show how we can add the constraint that $\mathbf{A}$ has to be invertible to the improved modelling and remove the *invalid* solutions without introducing new variables.

First, we take some equations from the system in Eq. (2) and use them to express $\mathbf{A}^{-1}$ in terms of $\mathbf{A}$. This is possible because the variables of $\mathbf{A}^{-1}$ appear only linearly and there are more that n^2 equations in the system. Specifically, we build the Macaulay matrix of the system, choosing an ordering such that the linear $\mathbf{A}^{-1}$-variables correspond to the leading columns. Then, we find the reduced row echelon form and take the first n^2 equations. They all contain only one linear $\mathbf{A}^{-1}$-variable, so the variable can be expressed as a quadratic equation in $\mathbf{A}$-variables. We use these terms to substitute the $\mathbf{A}^{-1}$-variables in the system corresponding to $\mathbf{AA}^{-1} = \mathbf{I}$. This approach yields $n^2 - n$ homogeneous and n inhomogeneous cubic equations in the $\mathbf{A}$-variables, that we add to the system derived from Eq. (3).

Since the new equations are all cubic, they do not influence greatly the asymptotic complexity of solving the system using a Gröbner basis algorithm like F4 [Fau99]. However, they are useful for eliminating the invalid solutions and they improve the running times for practical sizes. Hence, we use these equations in our experimental work, but we do not consider them in the complexity analysis in Sect. 6, or rather, we assume that they can only improve the solving complexity. It is commonly known that adding equations improves the solving time of Gröbner basis algorithms, and our experiments (in Sect. 7) show that this holds true for our case. In conclusion, we consider the following complexity analysis to be an upper bound, and, asymptotically, we do not expect it to differ a lot from the complexity analysis that includes the added cubic equations.

6 Complexity Analysis

The system obtained from Eq. (3) is a quadratic system of $n \cdot (\binom{n}{2} - n) = n^2 \cdot \frac{n-3}{2}$ equations in n^2 variables. With the assumption that this system is semi-regular, the asymptotic behavior of the degree of regularity can be estimated using [BFSY05]. Then, with $\alpha = \frac{n-3}{2}$, the resulting degree of regularity would grow as $d_{reg} \sim \frac{n}{4}$. However, as we will shortly see, the system is not semi-regular.

6.1 Non-trivial Syzygies

The exterior powers described in Sect. 2 hold a lot of extra structure. These will allow us to find extra syzygies in our system. Consider the following vector space:

$$L(\phi) := \{\omega \in \bigwedge^2 \mathbb{F}_q^n \mid \phi(\omega, z) = 0, \forall z \in \mathbb{F}_q^n\}.$$

This vector space can also be realized as the kernel of the following map:

$$\bigwedge^2 \mathbb{F}_q^n \to \mathbb{F}_q^n, \quad \omega \mapsto \begin{bmatrix} \phi(\omega, e_1) \\ \vdots \\ \phi(\omega, e_n) \end{bmatrix}.$$

This vector space enables a different perspective on the improved matrix-code modelling. The system described in Eq. (3) is also generated by

$$\{\phi_i(\mathbf{A}\omega) \mid 1 \le i \le n, \ \forall \omega \in L(\psi)\}. \quad (4)$$

Let us now consider the degree-3 elements of the ideal generated by the system above. This is a vector space generated by elements $\{a_{jk} \cdot \phi_i(\mathbf{A}\omega)\}$. For any combination (ω, i) there is a specific linear combination given by

$$\sum_j a_{ji}\phi_j(\mathbf{A}\omega) = \phi(\mathbf{A}\omega, \mathbf{A}\mathbf{e}_i) = \phi(\mathbf{A}(\omega \wedge \mathbf{e}_i)).$$

These linear combinations are all of the form $\phi(\mathbf{A}\theta)$ where $\theta \in \bigwedge^3 \mathbb{F}_q^n$. More specifically, $0 = \phi(\mathbf{A}\theta) = \psi(\theta)$, must hold for every θ, therefore $\theta \in \ker(\psi)$. With this structure in consideration let us look at the map

$$\xi_\psi : L(\psi) \otimes \mathbb{F}_q^n \to \ker(\psi), \quad (\omega, \mathbf{x}) \mapsto \omega \wedge \mathbf{x} \tag{5}$$

Of special interest are elements in the kernel of ξ_ψ. Let $\sum_k \omega_k \otimes \mathbf{e}_{i_k} \in \ker \xi_\psi$ then

$$\begin{aligned}
\sum_k \sum_j a_{ji_k}\phi_j(\mathbf{A}\omega) &= \sum_k \phi(\mathbf{A}\omega_k, \mathbf{A}\mathbf{e}_{i_k}) \\
&= \phi\left(\mathbf{A}\left(\sum_k \omega_k \wedge \mathbf{e}_{i_k}\right)\right) \\
&= \phi(\mathbf{A}(\mathbf{0})) \\
&\equiv 0.
\end{aligned}$$

Thus, we get a syzygy for each vector in the kernel of ξ_ψ. Let us call these *wedge syzygies.*

Remark 1. Empirical analysis for n up to 25 shows that this map is surjective for $n \in \{4, 5\} \cup \{7, \ldots, 25\}$ for random alternating trilinear forms. In the case $n = 6$, the image consistently has dimension one lower than $\ker(\psi)$. This might be interesting to look at from a mathematical point of view. However, for practical considerations we treat this as just a curiosity.

Now using the rank-nullity theorem we obtain the dimension for the module in degree 3 generated by wedge syzygies:

$$\left(\binom{n}{2} - n\right) \cdot n - \left(\binom{n}{3} - 1\right) = \frac{(n+1)(n-1)(n-3)}{3}.$$

6.2 Hilbert Series and the Solving Degree

We analyze how the system behaves under the block Wiedemann XL algorithm [Cop94]. For this we need the Hilbert series, the generating function for the monomials, and the density of our equations. In order to state the Hilbert series we have to make an assumption about the syzygies appearing in our system.

Assumption 1. *The syzygy module of the ideal in the system in Eq.* (4) *is generated by the trivial syzygies and the wedge syzygies.*

Using this assumption we can state the Hilbert series for the ideal generated by our system. To sum-up, we have a system of $\frac{n^2(n-3)}{2}$ quadratic equations in n^2 variables with $\frac{(n+1)(n-1)(n-3)}{3}$ syzygies in degree 3. First let us give the generating function for the amount of monomials in each degree as:

$$\mathcal{M}(t) = \frac{1}{(1-t)^{n^2}}.$$

Here we denote by $[t^\alpha]\mathcal{M}$ the coefficient of t^α in the series. Now we can state the Hilbert series:

$$\mathcal{H}(t) = (1-t^2)^{\frac{n^2(n-3)}{2}}(1-t^3)^{-\frac{(n+1)(n-1)(n-3)}{3}}\mathcal{M}(t).$$

Next let us look at the density of the equations in our system. In the modelling in Eq. (3) we take the product of the matrices $\mathbf{G}'^{\perp}$ and $\mathbf{G}(\mathbf{A}\otimes\mathbf{A})$. The dual code of $\mathcal{D}$ is of dimension $\binom{n}{2}-n$ in a vector space of dimension $\binom{n}{2}$. Therefore it can be represented by a basis of skew-symmetric matrices with $n+1$ non-zero entries in the upper-half triangle. Then taking the systematic form of $\mathbf{G}'^{\perp}$, we obtain $2(n+1)$ nonzero entries per row. On the other hand $\mathbf{G}(\mathbf{A}\otimes\mathbf{A})$ has a linear combination of $\binom{n-1}{2}$ terms $a_{ij}a_{i'j'}$ in every cell. Therefore, the density per equation is at most $2(n+1)\binom{n-1}{2}$.

The complexity for using the block Wiedemann XL algorithm is given by:

$$\mathcal{O}\left(\min_{\alpha,[t^\alpha]\mathcal{H}\leq 0} 3\cdot(n-2)(n-1)(n+1)\cdot([t^\alpha]\mathcal{M})^2\right).$$

Here the factor $(n-2)(n-1)(n+1)$ is the density and 3 is a hidden constant of the algorithm itself. Now a simple computation will give us the witness degree and complexities for solving ATFE systems. These are summarized in table Table 2.

7 Experimental Results

To confirm our theoretical findings, we implemented both the direct modelling described in Sect. 5.1 and the improved modelling with our proposed variant described in Sect. 5.3. Using this implementation, we perform experiments to confirm the estimates in our complexity analysis. In addition, we solve random instances of both modellings to compare the running times.

7.1 Computing Syzygies

In order to find the structure of the system of equations, we ran experiments to look for syzygies. This was done in two ways. In the first setting, we ran experiments by computing the entire Macaulay matrix up to certain degrees. However, since these experiments are computationally heavy we considered also another approach to be able to tell something for higher n. In the second setting, we looked at the kernel of ξ_ψ as in Eq. (5), as these generate syzygies.

Table 2. Solving degrees and complexities for ATFE instances using the improved matrix-code modelling.

n	d_{wit}	$\log_2$ complexity
8	9	83
9	9	90
10	9	95
11	9	101
12	9	105
13	9	110
14	10	123
15	10	127
20	11	155
25	13	193
30	14	219
35	15	245
40	17	283

Using Macaulay Matrices. We ran experiments on computing the Macaulay matrices for several degrees and several values for n. For this, we first generate the system of equations from our modelling. Next, we multiply all equations by all monomials of the corresponding degrees. Then, we construct the Macaulay matrix from this and finally, we row reduce in order to find the left nullity. The left nullity will tell us the amount of syzygies in the corresponding degree. The predicted amount of syzygies in each degree can be calculated from the Hilbert series and correspond to the coefficients of the following series:

$$\mathcal{S}(t) = n\left(\binom{n}{2} - n\right)t^2 - \mathcal{M}(t) + \mathcal{H}(t).$$

For $d = 3, 4$ this corresponds to

$$[t^3]\mathcal{S} = \frac{(n+1)(n-1)(n-3)}{3}$$

and

$$[t^4]\mathcal{S} = n^2\frac{(n+1)(n-1)(n-3)}{3} + \binom{n\left(\binom{n}{2} - n\right)}{2}.$$

Note that the resources required to run these calculations are high and this limits the size of n and d in our setup. The results can be found in Table 3.

From the results, we conclude that we correctly predict the amount of syzygies in degree 3 for the n values that we tested (except for $n = 6$) and that we correctly predict $n = 7, d = 4$. As we can see, for $n = 5, 6$, the predictions for $d = 4$ are off and extra syzygies appear. This is not surprising as for $n = 5$ and

Table 3. Experimental syzygies.

n	$d = 3$		$d = 4$	
	experiment	prediction	experiment	prediction
5	16	16	906	700
6	72	35	4149	2691
7	64	64	7889	7889
8	105	105		
9	160	160		
10	231	231		
11	320	320		

$n = 6$ we know that the automorphism groups are non-trivial. Furthermore, the matrices ϕ_i are of rank at most 4 since they have to be even and at most $n - 1$. This might also lead to extra syzygies in degree 4 for these two values of n. The fact that $n = 7$ is a correct prediction should give us some reassurance for higher n.

The Function ξ_ψ. Recall the function ξ_ψ that we introduced in Eq. (5). Since every element in $\ker(\xi_\psi)$ leads to a syzygy in degree three, it is worthwhile to explore its size. Then we can give a lower bound on the amount of syzygies. As stated before, we used experiments to verify that this is surjective for random alternating trilinear forms for n up to 30 (except $n = 6$). For each of those we computed the vector space $L(\psi)$. Then we created a list of wedge products of $\omega \in L(\psi)$ and canonical basis vectors $\mathbf{e}_i$. This results in a list of elements from $\bigwedge^3 \mathbb{F}_q^n$. These are just 3-way arrays so we vectorized them to vectors of length n^3. Finally, we computed the dimension of the space spanned by these vectors and verified this is the same dimension as $\ker(\psi)$. We conclude that the functions ξ_ψ are surjective for all these random instances and assume this holds for the generic case.

7.2 Running Gröbner Basis Computations

As a final step in our experimental work, we solve concrete instances of the systems arising from the quadratic with inverse modelling from [BDN+23] and the improved modelling, using the F4 [Fau99] implementation in MAGMA [BCP97]. For parameter sizes $n = \{5, 6, 7\}$, we generate 50 random instances of ATFE with one planted solution. We do this by generating a random trilinear form ϕ and a random invertible matrix $\mathbf{A}$, and then applying the group action to compute ψ. Note that for these parameter sizes ($n < 9$) we expect to have many solutions to the systems, so instead of enumerating the solution space, we stop after the computation of the Gröbner basis.

Table 4. Running times (in seconds) of F4 using two modellings of ATFE.

n	Modelling in [BDN+23]	Our modelling
5	64.20	0.64
6	> 200000	679.46

Results shown in Table 4 are an average of 50 runs. All of the instances are over $\mathbb{F}_q$ with $q = 3$, however, we performed (fewer instances of) these experiments with $q = 31$ and obtained comparable results. We see that the improved modelling significantly outperforms the quadratic with inverse modelling, which is in line with our theoretical findings. For $n = 7$, the computation for both variants timed out after 72 h. For $n = 6$, we were only able to solve the systems using the improved modelling. However, the authors of [BDN+23] report that they were able to solve the system for $n = 6$ in about 25 h with the quadratic with inverse modelling.

Acknowledgements. This research is supported by the NWO grant OCNW.M.21. 193 (ALPaQCa) and the ERC Starting Grant 805031 (EPOQUE).

References

[ADFMP20] Alamati, N., De Feo, L., Montgomery, H., Patranabis, S.: Cryptographic group actions and applications. In: Moriai, S., Wang, H. (eds.) ASIACRYPT 2020. LNCS, vol. 12492, pp. 411–439. Springer, Cham (2020). https://doi.org/10.1007/978-3-030-64834-3_14

[Bar04] Bardet, M.: Étude des systèmes algébriques surdéterminés. Applications aux codes correcteurs et à la cryptographie. Ph.D. thesis, Université de Paris VI (2004)

[BBC+20] Bardet, M., et al.: Improvements of algebraic attacks for solving the rank decoding and MinRank problems. In: Moriai, S., Wang, H. (eds.) ASIACRYPT 2020. LNCS, vol. 12491, pp. 507–536. Springer, Cham (2020). https://doi.org/10.1007/978-3-030-64837-4_17

[BBPS21] Barenghi, A., Biasse, J.-F., Persichetti, E., Santini, P.: LESS-FM: fine-tuning signatures from the code equivalence problem. In: Cheon, J.H., Tillich, J.-P. (eds.) PQCrypto 2021 2021. LNCS, vol. 12841, pp. 23–43. Springer, Cham (2021). https://doi.org/10.1007/978-3-030-81293-5_2

[BCP97] Bosma, W., Cannon, J., Playoust, C.: The magma algebra system. I. The user language. J. Symbolic Comput. **24**(3–4), 235–265 (1997). Computational algebra and number theory (London, 1993)

[BDN+23] Bläser, M., et al.: The ALTEQ Signature Scheme: Algorithm Specifications and Supporting Documentation. NIST PQC Submission (2023)

[Beu20] Beullens, W.: Not enough LESS: an improved algorithm for solving code equivalence problems over $\mathbb{F}_q$. In: Dunkelman, O., Jacobson, Jr., M.J., O'Flynn, C. (eds.) SAC 2020. LNCS, vol. 12804, pp. 387–403. Springer, Cham (2021). https://doi.org/10.1007/978-3-030-81652-0_15

[Beu22] Beullens, W.: Graph-theoretic algorithms for the alternating trilinear form equivalence problem. Cryptology ePrint Archive, Paper 2022/1528 (2022). https://eprint.iacr.org/2022/1528

[BFFP11] Bouillaguet, C., Faugère, J.-C., Fouque, P.-A., Perret, L.: Practical cryptanalysis of the identification scheme based on the isomorphism of polynomial with one secret problem. In: Catalano, D., Fazio, N., Gennaro, R., Nicolosi, A. (eds.) PKC 2011. LNCS, vol. 6571, pp. 473–493. Springer, Heidelberg (2011). https://doi.org/10.1007/978-3-642-19379-8_29

[BFSY05] Bardet, M., Faugère, J.-C., Salvy, B., Yang, B.-Y.: Asymptotic behaviour of the degree of regularity of semi-regular polynomial systems. In: Proceedings of MEGA 2005, Eighth International Symposium on Effective Methods in Algebraic Geometry (2005)

[BFV13] Bouillaguet, C., Fouque, P.-A., Véber, A.: Graph-theoretic algorithms for the "isomorphism of polynomials" problem. In: Johansson, T., Nguyen, P.Q. (eds.) EUROCRYPT 2013. LNCS, vol. 7881, pp. 211–227. Springer, Heidelberg (2013). https://doi.org/10.1007/978-3-642-38348-9_13

[BKP20] Beullens, W., Katsumata, S., Pintore, F.: Calamari and Falafl: logarithmic (linkable) ring signatures from isogenies and lattices. In: Moriai, S., Wang, H. (eds.) ASIACRYPT 2020. LNCS, vol. 12492, pp. 464–492. Springer, Cham (2020). https://doi.org/10.1007/978-3-030-64834-3_16

[BMPS20] Biasse, J.-F., Micheli, G., Persichetti, E., Santini, P.: LESS is more: code-based signatures without syndromes. In: Nitaj, A., Youssef, A. (eds.) AFRICACRYPT 2020. LNCS, vol. 12174, pp. 45–65. Springer, Cham (2020). https://doi.org/10.1007/978-3-030-51938-4_3

[Bou11] Bouillaguet, C.: Algorithms for some hard problems and cryptographic attacks against specific cryptographic primitives. Ph.D. thesis, Université Paris Diderot (2011)

[CNP+22] Chou, T., et al.: Take your meds: digital signatures from matrix code equivalence. Cryptology ePrint Archive, Paper 2022/1559 (2022). https://eprint.iacr.org/2022/1559

[Cop94] Coppersmith, D.: Solving homogeneous linear equations over gf (2) via block wiedemann algorithm. Math. Comput. **62**, 333–350 (1994)

[Cou06] Couveignes, J.-M.: Hard homogeneous spaces. Cryptology ePrint Archive, Paper 2006/291 (2006)

[DFG19] De Feo, L., Galbraith, S.D.: SeaSign: compact isogeny signatures from class group actions. In: Ishai, Y., Rijmen, V. (eds.) EUROCRYPT 2019. LNCS, vol. 11478, pp. 759–789. Springer, Cham (2019). https://doi.org/10.1007/978-3-030-17659-4_26

[Fau99] Faugère, J.-C.: A new efficient algorithm for computing gröbner basis (F4). J. Pure Appl. Algebra **139**(1–3), 61–88 (1999)

[FdVP08] Faugère, J.-C., Levy-dit-Vehel, F., Perret, L.: Cryptanalysis of MinRank. In: Wagner, D. (ed.) CRYPTO 2008. LNCS, vol. 5157, pp. 280–296. Springer, Heidelberg (2008). https://doi.org/10.1007/978-3-540-85174-5_16

[FP06] Faugère, J.-C., Perret, L.: Polynomial equivalence problems: algorithmic and theoretical aspects. In: Vaudenay, S. (ed.) EUROCRYPT 2006. LNCS, vol. 4004, pp. 30–47. Springer, Heidelberg (2006). https://doi.org/10.1007/11761679_3

[GQ21] Grochow, J.A., Qiao, Y.: On the complexity of isomorphism problems for tensors, groups, and polynomials I: tensor isomorphism-completeness. In:

Lee, J.R. (ed.)12th Innovations in Theoretical Computer Science Conference (ITCS 2021), volume 185 of Leibniz International Proceedings in Informatics (LIPIcs), pp. 31:1–31:19, Dagstuhl, Germany, 2021. Schloss Dagstuhl-Leibniz-Zentrum für Informatik (2021)

[Gre12] Greub, W.H.: Multilinear Algebra. Grundlehren der mathematischen Wissenschaften. Springer, Heidelberg (2012)

[Leo82] Leon, J.S.: Computing automorphism groups of error-correcting codes. IEEE Trans. Inf. Theory **28**(3), 496–510 (1982)

[Pat96] Patarin, J.: Hidden Fields Equations (HFE) and Isomorphisms of Polynomials (IP): two new families of asymmetric algorithms. In: Maurer, U. (ed.) EUROCRYPT 1996. LNCS, vol. 1070, pp. 33–48. Springer, Heidelberg (1996). https://doi.org/10.1007/3-540-68339-9_4

[Per05] Perret, L.: A fast cryptanalysis of the isomorphism of polynomials with one secret problem. In: Cramer, R. (ed.) EUROCRYPT 2005. LNCS, vol. 3494, pp. 354–370. Springer, Heidelberg (2005). https://doi.org/10.1007/11426639_21

[RS06] Rostovtsev, A., Stolbunov, A.: Public-key cryptosystem based on isogenies. Cryptology ePrint Archive, Paper 2006/145 (2006)

[RST22] Reijnders, K., Samardjiska, S., Trimoska, M.: Hardness estimates of the code equivalence problem in the rank metric. Cryptology ePrint Archive, Paper 2022/276 (2022)

[TDJ+22] Tang, G., Duong, D.H., Joux, A., Plantard, T., Qiao, Y., Susilo, W.: Practical post-quantum signature schemes from isomorphism problems of trilinear forms. In: Dunkelman, O., Dziembowski, S. (eds.) EUROCRYPT 2022. LNCS, vol. 13277, pp. 582–612. Springer, Cham (2022). https://doi.org/10.1007/978-3-031-07082-2_21

Modeling Noise-Accepting Key Exchange

Elsie Mestl Fondevik[1,2(✉)] and Kristian Gjøsteen[1]

[1] Norwegian University of Science and Technology, Trondheim, Norway
{elsie.mestl,kristian.gjosteen}@ntnu.no
[2] Kongsberg Defence & Aerospace, Kongsberg, Norway

Abstract. In this paper we use code-based public key encryption schemes to construct key exchange protocols that are suitable for use in the presence of transmission bit-errors, such as in mobile ad hoc networks.

Building upon the security model by Bellare and Rogaway [2], we let instances that have matching conversations up to a certain error bound generate the same session key. In order to prevent an adversary from trivially attacking the schemes, a relaxed version of matching conversation is introduced and shown to be well-defined.

To give validity to our model we show that the introduced security model can be reduced to the original Bellare and Rogaway-security model. Additionally, we prove the naive and obvious solution of adding error correction to a key exchange protocol will not affect the security of the protocol.

Finally, we introduce the concept of error-resistant asymmetric schemes and key encapsulations. Then through a modified Fujisaki-Okamoto-transform we show that a probabilistic error-resistant asymmetric scheme can be transformed into an error-resistant key encapsulation mechanism (KEM). A key exchange protocol construction based on the transformed KEM's is then presented and proven secure.

Keywords: Key Exchange · Unreliable Networks · Error Correction · Emergency Networks

1 Introduction

In the key exchange security model by Bellare and Rogaway [2], it is a fundamental requirement that two instances with distinct transcripts should not produce the same session key with more than a negligible probability. If this property is not preserved then this may give rise to the following trivial attack: The adversary chooses two instances with distinct transcripts; challenging one while revealing the session key of the other. If the protocol is as described then the adversary can select the two instances such that there is a non-negligible probability that they produce the same session key.

A. Esser and P. Santini (Eds.): CBCrypto 2023, LNCS 14311, pp. 104–124, 2023.
https://doi.org/10.1007/978-3-031-46495-9_6

For everyday network usage (Internet access and WiFi connection), where communication happens over e.g. a TCP connection[1] [4], the perfect transmission requirement does not introduce problems. For other networks, such as mobile ad hoc networks (MANET) used in emergency networks by search and rescue personnel [12,15], TCP is not a suitable protocol due to the expected amount of errors on the network which drops TCP traffic. Bandwidth constraints and availability requirements prevent additional error correction to be added or increased. Instead UDP without checksum [14] may be utilized, to prevent flooding the network with re-transmitted packets, to accept packets containing errors. If the network is mainly used for voice or video streams and a symmetric key has already been established this is a manageble problem. This is because some block cipher modes have no error propagation, i.e. single bit-errors in the ciphertext translate to single bit-errors in the message. Asymmetric encryption or key exchange, on the other hand, does in general not have this feature. The current solution for key exchange and distribution in such networks is to use pre-shared symmetric keys to obtain the desired security goals [17]. We want to improve on this.

1.1 Our Contributions

In this paper, we build upon the security model of Bellare and Rogaway [2] to gain a security notion that allows some noise on message transcripts. Furthermore, we use a code-based public key encryption scheme to construct a key exchange protocol that is suitable for use in the presence of bit-errors that occur during transmission.

We introduce a lower bound, *good distance bound* e, specifying the minimum amount of noise that an error-resistant key exchange protocol must tolerate. Meaning, any instance with transcript less than e apart from the noiseless original transcript must generate the intended session key. Since different unreliable networks have different error distributions and expected error rates, both the value for good distance bound and the metric used to determine the quantity of noise is incorporated into the protocol design choices.

The adversary is still allowed full network control, meaning it is in control of all the noise added to transmitted messages. The ideal transcripts will therefore not be known to the individual instances nor the simulator running the experiment, making computing the difference between a received transcript and the ideal transcript an impossible task for anyone but the adversary. This gives the adversary an avenue for the trivial attack above, and, although meant to be prevented, cannot be observed since the information needed to stop the adversary is missing. To solve this we remove the requirement specifying that two instances with matching conversations must generate the same session key. Instead we say that two instances are *loosely related* if they have transcripts at most $2e$ apart. Thus preventing the adversary from breaking freshness in a way that is perceivable by the experiment.

[1] TCP requires that received packets are error-free. If a packet containing error is received by the intended recipient over TCP connection a re-transmittal of the specific data packet will be requested.

In order to prove the validity of our model we show that by setting the parameters appropriately the error-resistant security model is equivalent to the original model by Bellare and Rogaway [2]. We also show that the naive and obvious solution of adding error correction to a key exchange protocol will result in a protocol in the modified model that is as secure as the original protocol.

The final part of the paper concerns a protocol specifically designed for a noisy environment. By assuming the existence of an asymmetric encryption scheme that will decrypt to the correct message even if the ciphertext contains errors, we develop a modified Fujisaki-Okamoto-transform that preserves this error-resistant property when transforming the asymmetric scheme into a key encapsulation mechanism (KEM). The KEM can then be applied in a general construction of a KEM-based two-message key exchange protocol with the additional error resistance inherited from the KEMs. The existence of an error-resistant asymmetric encryption scheme is easily shown by choosing suitable parameters of selected code-based schemes like McEliece [13].

1.2 Related Work

As discussed in the previous section the key exchange model presented by Bellare and Rogaway (BR) [2] is not suitable for noisy network channels. Either opening for trivial attacks or preventing a correct session key from being established.

The modified BR model presented by Li and Schäge [11] removes the requirement of matching conversation to determine partnership and instead considers the produced session key in order to determine partnering relations. Although this partnering relation would include our model, it is a highly generalized definition and does not specify when transcripts should or should not lead to identical session keys, as done in our model. Additional overlaying restrictions would also be required to encompass concepts such as noise variation. As a result we have decided to build directly upon Bellare and Rogaway [2]. This choice also gives us the benefit of a publicly computable (loose) partnering relation.

Error correcting codes focus on sending messages error-free over networks, but do not offer security [7]. Some work has been put into combining error correction codes and encryption [5,16]. They focus, however, on symmetric encryption and not key exchange as done here. Furthermore, they do not aim to present a refined security model for unreliable networks, but rather present a specific protocol. As such, CryptCode [5], when modified for key exchange, can be considered a use-case for our generalized model, as seen in Theorem 3.

Finally, fully homomorphic encryption has recently introduced a notion of approximate fully homomorphic encryption that allow approximate decryption [10], i.e. decryptions that are correct up to a certain error margin. This can be considered the dual problem of what we wish to achieve; correct decryptions on ciphertext with some small amount of error.

2 Prerequisites

Informally, key exchange is an interaction between two entities that, hopefully, results in the generation of a *session key* k. The participating parties are called *instances* and have either an initiating or a responding role in the exchange. In order to differentiate between instances they will be numbered as they occur. The enumeration has no other function than to improve readability and help with bookkeeping.

Each instance has access to a long term key pair, $(\mathsf{pk}, \mathsf{sk})$, consisting of a public key pk and a secret key sk. Multiple instances may share the same long term key. Public information relating to the environment and the context of a key exchange protocol execution is called *associated data, ad*.

In the following definition, Definition 1, we formally define key exchange. The definition is based on Bellare and Rogaway [2].

Definition 1. *A two-party key exchange protocol* $\text{KEX} = (\mathfrak{K}, \mathcal{K}, \mathcal{I}, \mathcal{R})$ *consists of a set of keys* $\mathfrak{K}$ *and the following three algorithms:*

- *The* key generation algorithm $\mathcal{K}$ *takes no argument as input and outputs a pair of long-term keys* $(\mathsf{pk}, \mathsf{sk})$.
- *The* initiator algorithm $\mathcal{I}$ *takes associated data, ad, long-term key pair,* $(\mathsf{sk}_{\mathcal{I}}, \mathsf{pk}_{\mathcal{I}})$ *and a public key,* $\mathsf{pk}_{\mathcal{R}}$*, as input. It exchanges multiple messages with the responder* $\mathcal{R}$ *before outputting either a session key* k *or* $\perp$ *to signify failure.*
- *The* responder algorithm $\mathcal{R}$ *takes associated data, ad, long-term key pair,* $(\mathsf{sk}_{\mathcal{R}}, \mathsf{pk}_{\mathcal{R}})$*, and a public key,* $\mathsf{pk}_{\mathcal{I}}$*, as input. It exchanges multiple messages with the initiator* $\mathcal{I}$ *before outputting either a session key* k *or* $\perp$ *to signify failure.*

The transcript, tr*, of an initiator or responder consists of the sequence of messages sent and received by the algorithm together with the associated data and public keys. If both* $\mathcal{I}(ad, \mathsf{sk}_{\mathcal{I}}, \mathsf{pk}_{\mathcal{I}}, \mathsf{pk}_{\mathcal{R}})$ *and* $\mathcal{R}(ad, \mathsf{sk}_{\mathcal{R}}, \mathsf{pk}_{\mathcal{R}}, \mathsf{pk}_{\mathcal{I}})$ *output keys when run using identical transcript* tr *the keys will be identical.*

Two instances are considered partners if they have *matching conversations*, i.e. they have identical transcripts. Matching conversation forms a reflexive relation between instances, denoted $\sim$. Any two instances i, j where $i \sim j$ will produce the same session key. An instance is *implicitly authenticated* if it has at most one partner and that partner should be of the opposite role.

The security of the model is game based, and runs as an experiment hosted by a simulator, see Fig. 1. The adversary controls the experiment though a series of queries to the simulator using public information. At some point during run time the adversary may issue a *real or random (ROR)* challenge to an instance. The goal of the adversary is to determine whether the returned key is a real or fake session key. The adversary wins if it determines correctly.

The adversary may query the simulator to reveal information about specific instances in order to improve its success rate. An instance is *exposed* if one of the

$\mathsf{Exp}^{\mathsf{ROR}}_{\mathbf{KEX}}(\mathcal{A})$

$b, b'' \leftarrow\!\!\$ \{0,1\}$
$b' \leftarrow\!\!\$ \mathcal{A}^{\mathsf{ROR}}_{\mathbf{KEX}}()$
Stop with b'

KeyGen():

The j'th key gen query is initiated
Compute $(\mathsf{sk}, \mathsf{pk}) \leftarrow \mathcal{K}$
Record $(j, \mathsf{sk}, \mathsf{pk})$
Give pk to $\mathcal{A}$

Execute($\rho, j, \mathsf{pk}', ad$):

The i'th execute query is initiated
If $(j, \mathsf{sk}, \mathsf{pk})$ not recorded:
  return $\perp$
If $\rho = \mathcal{I}$:
  Start $\mathcal{I}(ad, \mathsf{sk}, \mathsf{pk}, \mathsf{pk}')$
  If instance sends message m:
    Give (i, m) to $\mathcal{A}$
If $\rho = \mathcal{R}$:
  Start $\mathcal{R}(ad, \mathsf{sk}, \mathsf{pk}, \mathsf{pk}')$
  Give i to $\mathcal{A}$

Long-Term Key Reveal(j):

If $(j, \mathsf{sk}, \mathsf{pk})$ is recorded:
  Give sk to $\mathcal{A}$

Send(i, m):

If (i, s_0, s_1) is recorded:
  Give $\perp$ to $\mathcal{A}$
  return
Else:
  Send m to i'th instance
If the instance outputs $\perp$:
  Record $(i, \perp, \perp)$
  Give $(i, \perp)$ to $\mathcal{A}$
Else if instance outputs $\mathsf{k}_0 \in \mathfrak{K}$:
  $\mathsf{k}_1 \leftarrow\!\!\$ \mathfrak{K}$
  Record $(i, \mathsf{k}_0, \mathsf{k}_1)$
Else if instance sends message m':
  Give (i, m') to $\mathcal{A}$
Else:
  Give $(i, \top)$ to $\mathcal{A}$

Challenge(i):

If (i, s_0, s_1) is recorded:
  Give s_b to $\mathcal{A}$

Session Key Reveal(i):

If (i, s_0, s_1) is recorded:
  Give s_0 to $\mathcal{A}$

State Reveal(i):

Give random tape of i to $\mathcal{A}$

Fig. 1. A *real-or-random (ROR)* experiment game $\mathsf{Exp}^{\mathsf{ROR}}_{\mathrm{KEX}}(\mathcal{A})$ for key exchange KEX $= (\mathfrak{K}, \mathcal{K}, \mathcal{I}, \mathcal{R})$.

following query combinations has been issued: (1) a session key reveal query has been sent to itself or its partnered instance; (2) both a long-term key reveal and state reveal query has been sent to itself or its partnered instance; (3) a challenge query has been sent to both the instance and its partner. An instance that is not exposed is *fresh*. Challenges should only be issued towards fresh instances.

The formal ROR security definition for key exchange is given in Definition 2.

Definition 2. *Let* KEX *be a key exchange protocol, let* $\sim$ *be a partnering relation and let* $\mathcal{A}$ *be an adversary against* KEX *that can perform at most* l_p *key generation queries,* l_t *test queries and* l_i *execute queries.*

Let $\mathsf{E_a}$ *be the event that implicit authentication is broken and* $\mathsf{E_d}$ *be the event that the experiment* $\mathsf{Exp}^{\mathsf{ROR}}_{\Xi}(\mathcal{A})$ *terminates with the correct output* b *without breaking freshness, or the random bit* b'' *when breaking freshness. The advantage of* $\mathcal{A}$ *is then defined as*

$$\mathsf{Adv}^{\mathsf{ror}}_{\mathrm{KEX}}(\mathcal{A}) = \max\left\{2\left|\Pr[\mathsf{E_d}] - \tfrac{1}{2}\right|, \Pr[\mathsf{E_a}]\right\}.$$

3 Key Exchange over Unreliable Networks

Unreliable networks are networks where there is a high probability of packet loss, noise and varying degree of bandwidth and connectivity. Data sent over such networks may, therefore, be received out of order, partially lost or distorted. We aim to develop a model for key exchange protocols that can function in such environments.

3.1 Modeling Noise

In order to modify the security model for key exchange to fit unreliable networks we need to decide how noise is added to messages in the security game. There are two choices; either the noise is added by the experiment or by the adversary.

If we require that the security experiment adds the noise we may gain a more realistic model as the adversary will often be located somewhere on the network and would only see noisy messages. It is important to point out that having the security experiment be in charge of adding noise does not prevent the adversary from successfully adding additional noise in such a way that identical session keys are generated. On the other hand, allowing the adversary to control all added noise is a stronger requirement. Not only will the adversary see all transmitted messages without noise, but it also has full control over how much noise is added before passing the message to its intended recipient. As there should be no minimum amount of noise required to use protocols for unreliable networks, i.e. they should be secure in the standard noise-free environment, we will use adversarially controlled noise in our model. Similarly, we do not require the adversary to add noise according to a specified distribution.

Due to the noise added to messages sent over unreliable networks we distinguish between sent and received messages. We define *sent messages* as messages sent from an instance onto the network. The *received messages* are delivered from the network to an instance. Sent messages are always as generated by the instance, while received messages may contain errors. With adversarial-controlled noise the adversary will always know the sent message and decides how the received message is altered. A noise free transcript, that is a transcript that consist only of sent messages, is called an *ideal transcript.*

The noise added to messages in an unreliable network happens during message transmission. As all messages are encoded as bit-streams, any noise on the message can be interpreted as bit-flips. However, the transmission usually contain error correction on lower levels of the OSI-model [3], even when working over unreliable networks. Therefore, depending on the error correction used by the network, the noise distribution on the received messages may not be uniform. Protocols should consequently be designed to match the expected noise distribution of the specific network it is intended for. As such, when differentiating between sent and received messages, depending on the protocol design, it might not be enough to just calculate the Hamming distance. A specific metric should in other words be determined during protocol construction. Similarly, the lower bound of how far apart a transcript can be from the ideal transcript and still generate the correct session key, should also be protocol-specific.

Definition 3. *Let d be a metric and let $e > 0$. A two-party key exchange protocol* KEX *is e-error-resistant over d if any instance i with transcript tr' produces the correct session key if $d(\mathsf{tr}, \mathsf{tr}') \leq e$ where tr is an ideal transcript.*
The value e is called a good distance bound.

The metric d and the good distance bound e refer to the pattern of errors and the "amount" of errors an e-error-resistant protocol over d is able to withstand. We point out that just because an instance can process a protocol-specific transcript containing some noise does not mean that an instance is able to interpret any transcript or message sequence. For instance the sent and received messages need to be of the correct length and format. Additionally, the public keys and associated data needs to be as expected, i.e. the transcript needs to be in accordance to protocol. The latter requirement will ensure that error-resistant key exchange protocols are no more susceptible too man in the middle attacks than their noise free counterparts.

3.2 Noisy Matching Conversations

The natural next step now is to extend the matching conversation relation to tolerate noise. We say that two instances i and j of an e-error-resistant KEX over d have *e-matching conversation* if their ideal transcripts are identical. It is clear that e-matching conversations is a valid partnering relation for e-error-resistant key exchange protocols, as any two instances with identical transcripts will by Definition 1 generate the same session key. However, with the exception of the adversary it is hard for anyone, simulator included, to determine partnership using e-matching conversation since the ideal transcript is unknown, defeating the purpose of a partnering relation.

From triangular inequality property of metrics we know that if two instances i, j with transcripts $\mathsf{tr}_i, \mathsf{tr}_j$ have e-matching conversation, then $d(\mathsf{tr}_i, \mathsf{tr}_j) \leq 2e$. The opposite, however, does not necessarily hold; just because two transcripts happen to be less than $2e$ apart does not mean they generate the same session key, meaning it is not immediately straight forward to use the metric to determine partnership.

Remember the point of the matching conversation and other partnering relations is to prevent the adversary from trivially winning the game. It is used to calculate freshness so that an adversary cannot challenge and reveal information pertaining to the same session key. So, if we relax the requirement that partnered instances have to generate the same session key to only requiring that they generate the same session key with non-negligible probability, then we might be able to use d to say all instances with transcripts closer than some distance are *loosely related*. The natural value for this bound is $2e$. Note, that due to hash collisions and the general nature of the birthday paradox we cannot require that an adversary never will be able to find collisions outside of the ball $\mathcal{B}_{2e}$. The choice of d and e needs to be carefully selected to give maximum coverage and prevent transcripts outside of the $2e$-ball to generate the correct codeword with non-negligible probability.

Freshness can then be defined as in Sect. 2, substituting matching conversations for loosely related instances. We say that an instance that is fresh under this new notion is *e-fresh*, dually an exposed instance is *e-exposed*.

The requirement for implicit authentication remains that an instance i has no more than one partner of the opposite role, using loosely related as the partnering relation. However, we add the additional requirement that the partners all need to produce the correct session key in order for it to be a valid authentication break. We call this slightly modified authentication requirement *e-implicit authentication.*

As in the original security definition given in Definition 2, let $\mathsf{E_d}$ be the event that an adversary can determine whether it is presented a real or random key. If the adversary sends a test query to an e-exposed instance, the adversary will automatically lose. Since we use a freshness notion that uses a loose partnering relation, the event $\mathsf{E_d}$ might in some cases hold even when the same session key has not been computed. This is different from the original model, but will only give any adversary a higher advantage, hence it is not a problem when computing the security of a key exchange protocol.

To summarize, we allow the adversary to add some noise to messages when running the experiment without affecting key generation, freshness or authentication. The modifications presented in Definition 3 and freshness is sufficient, so we can continue using $\mathsf{Exp}^{\mathsf{ROR}}_{\mathrm{KEX}}$ from Fig. 1. The security definition for noisy key exchange now follows naturally.

Definition 4. *Let* KEX *be an e-error-resistant key exchange protocol with metric d, and let $\mathcal{A}$ be an adversary against* KEX *that can perform at most l_p key generation queries, l_t test queries and l_i execute queries.*

Let $\mathsf{E_a}$ be the event that e-implicit authentication is broken and $\mathsf{E_d}$ be the event that the experiment $\mathsf{Exp}^{\mathsf{ROR}}_{\Xi}(\mathcal{A})$ terminates with the correct output b without breaking e-freshness using metric d. The advantage of $\mathcal{A}$ is then defined as

$$\mathsf{Adv}^{\mathsf{ror}}_{e-\mathrm{KEX}}(\mathcal{A}) = \max\left\{2\left|\Pr[\mathsf{E_d}] - \tfrac{1}{2}\right|, \Pr[\mathsf{E_a}]\right\}.$$

4 The Validity of Our Model

To argue the validity of our model we need to show it fulfills specific requirements. First, the model has to be applicable for regular key exchange protocols in an ideal situation, i.e. a noiseless channel with zero error allowance. In this ideal setting the adversarial advantage given by our model and that given by the original Bellare and Rogaway [2] should be indistinguishable. Second, given any key exchange protocol secure under Bellare and Rogaway [2], by adding enough error correcting code the corresponding protocol should be secure under the introduced security model. In other words, we need to show that the intuitive and canonical way of generating a noise resistant key exchange protocol holds.

4.1 Secure in a Noise-Free Environment

It is clear that any error-resistant key exchange protocol where the good distance bound is zero does not tolerate noise. The 0-matching conversation requires that only transcripts identical to the ideal transcripts should generate the correct session key, and only instances with identical transcripts are considered loosely related. The strict equality requirements is identical to the original key exchange protocol from Sect. 2. Note that this holds true independently of choice of d.

4.2 Canonical Extension to a Noisy Environment

We construct a noise-resistant key exchange protocol in the canonical way using an error correcting code together with a noiseless key exchange protocol. The constructed protocol is then shown to be just as secure as the noiseless version.

The construction follows naturally by adding a layer of error correction to all messages before transmission. When the message is received it is first decoded before being processed. More formally, let $(\mathfrak{K}, \mathcal{K}, \mathcal{I}_0, \mathcal{R}_0)$ be any key exchange protocol for ideal situation, and let $(\mathcal{E}, \mathcal{D})$ be an error correcting code. For noisy channels we define $(\mathfrak{K}, \mathcal{K}, \mathcal{I}, \mathcal{R})$ where

$\mathcal{I}$: Uses $\mathcal{I}_0$ as an internal algorithm. Any message sent by $\mathcal{I}_0$ is first encoded using $\mathcal{E}$ before sent by $\mathcal{I}$. Any message received by $\mathcal{I}$ is decoded using $\mathcal{D}$ before given to $\mathcal{I}_0$. If $\mathcal{I}_0$, $\mathcal{E}$ or $\mathcal{D}$ at any point output $\perp$ then $\mathcal{I}$ stops and outputs $\perp$. Else, when $\mathcal{I}_0$ outputs k, $\mathcal{I}$ outputs similarly.

$\mathcal{R}$: Uses $\mathcal{R}_0$ as an internal algorithm. Any message received by $\mathcal{R}$ is decoded using $\mathcal{D}$ before given to $\mathcal{R}_0$. Any message sent by $\mathcal{R}_0$ is first encoded using $\mathcal{E}$ before sent by $\mathcal{R}$. If $\mathcal{R}_0$, $\mathcal{E}$ or $\mathcal{D}$ at any point output $\perp$ then $\mathcal{R}$ stops and outputs $\perp$. Else, when $\mathcal{R}_0$ outputs k, $\mathcal{R}$ outputs similarly.

To prove the construction is secure in the new model, we use the fact that we can freely determine the metric used. Let x, y be sent or received messages. We define d^* as follows

$$d^*(x, y) = \begin{cases} 0 & \text{if } x = y \\ 1 & \text{if } \mathcal{D}(x) = \mathcal{D}(y) \\ 3 & \text{otherwise} \end{cases} .$$

Given transcripts tr_x and tr_y consisting of messages $(x_0, x_1, x_2, \cdots, x_n)$ and $(y_0, y_1, y_2, \cdots, y_n)$, respectively, we expand d^* to d as follows:

$$d(\mathsf{tr}_x, \mathsf{tr}_y) = \max\{d^*(x_i, y_i) \,:\, i = 0, \cdots, n\}.$$

Then only if all sent and received messages in the two transcripts decode to the same messages, will d output either a zero or a one. By using the metric d together with good distance bound $e = 1$, we get that any two instances are loosely related if they have transcripts no further apart than 2, meaning they deviate from the same codeword. From here the following statement and reduction follows straight forward.

Theorem 1. *Let $(\mathcal{E}, \mathcal{D})$ be an error correcting code, let* KEX_0 *be a key exchange protocol for noiseless channels, let d be the metric described above, and let* KEX *be the key exchange protocol constructed above using* KEX_0 *and* $(\mathcal{E}, \mathcal{D})$.

Then KEX *is a 1-error-resistant key exchange protocol over d. Furthermore, there exists an (l_i, l_p, l_t)-adversary $\mathcal{A}$ against* KEX *that uses time at most t if and only if there exists an (l_i, l_p, l_t)-adversary $\mathcal{A}_0$ against* KEX_0 *that uses at most t' time, where t and t' are essentially equal and*

$$\mathsf{Adv}^{\mathsf{ror}}_{\mathrm{KEX}_0}(\mathcal{A}_0) = \mathsf{Adv}^{\mathsf{ror}}_{1-\mathrm{KEX}}(\mathcal{A}).$$

Proof. By construction and choice of d it follows directly from definition that KEX is a 1-error-resistant key exchange protocol.

The second part of the proof will be a two way reduction between $\mathcal{A}$ and $\mathcal{A}_0$. Given an adversary $\mathcal{A}_0$ against KEX_0 we create an adversary $\mathcal{A}$ against KEX.

$\mathcal{A}$: When $\mathcal{A}_0$ outputs b, $\mathcal{A}$ outputs b.

Key gen query: When $\mathcal{A}_0$ sends a key gen query to KEX_0, $\mathcal{A}$ send key gen query to KEX. When $\mathcal{A}$ receives public long term key pk from the oracle, it passes pk to $\mathcal{A}_0$.

Execute query: When $\mathcal{A}_0$ sends an execute query to KEX_0, $\mathcal{A}$ sends execute query to KEX. If $\mathcal{A}$ receives confirmation from oracle, this is passed along (if received) to $\mathcal{A}_0$.

Send query: When $\mathcal{A}_0$ sends a send query (i, m) to KEX_0, $\mathcal{A}$ sends a send query $(i, \mathcal{E}(m))$ to KEX. If $\mathcal{A}$ receives (i, c) it passes $(i, \mathcal{D}(c))$ along to $\mathcal{A}_0$.

Test query: If $\mathcal{A}_0$ sends test query to KEX_0, the constructed adversary $\mathcal{A}$ will use this as its challenge as well, passing the query to KEX. When $\mathcal{A}$ receives challenge (i, s_b) it forwards the challenge to $\mathcal{A}_0$.

Session key reveal: If $\mathcal{A}_0$ sends session reveal query to KEX_0, the query is passed to KEX. When $\mathcal{A}$ receives key (i, s_0) it forwards it to $\mathcal{A}_0$.

State reveal: If $\mathcal{A}_0$ sends state reveal query to KEX_0, the query is forwarded to KEX. When $\mathcal{A}$ receives random tape from KEX, it forwards this to $\mathcal{A}_0$.

Long term key reveal: If $\mathcal{A}_0$ sends long term reveal query to KEX_0, $\mathcal{A}$ passes query along to KEX. When $\mathcal{A}$ receives key pair $(\mathsf{pk}, \mathsf{sk})$, the key pair is forwarded to $\mathcal{A}_0$.

It is clear that if $\mathcal{A}_0$ outputs the correct b then so does $\mathcal{A}$. Furthermore, if a session is non-fresh in $\mathsf{Exp}^{\mathsf{ROR}}_{\mathrm{KEX}}(\mathcal{A})$ it has identical underlying sent and received messages, meaning it is fresh in $\mathsf{Exp}^{\mathsf{ROR}}_{\mathrm{KEX}_0}(\mathcal{A}_0)$. If $\mathcal{A}_0$ breaks implicit authentication, then $\mathcal{A}$ will as well. Thus we have that

$$\mathsf{Adv}^{\mathsf{ror}}_{1-\mathrm{KEX}}(\mathcal{A}) \geq \mathsf{Adv}^{\mathsf{ror}}_{\mathrm{KEX}_0}(\mathcal{A}_0).$$

Given an adversary $\mathcal{A}$ against KEX we create an adversary $\mathcal{A}_0$ against KEX_0.

$\mathcal{A}_0$: When $\mathcal{A}$ outputs b, $\mathcal{A}_0$ outputs b.

Key gen query: When $\mathcal{A}$ sends a key gen query to KEX, $\mathcal{A}_0$ send key gen query to KEX_0. When $\mathcal{A}_0$ receives public long term key pk from the oracle, it passes pk to $\mathcal{A}$.

Execute query: When $\mathcal{A}$ sends an execute query to KEX, $\mathcal{A}_0$ sends execute query to KEX_0. If $\mathcal{A}_0$ receives confirmation from oracle, this is passed along (if received) to $\mathcal{A}$.

Send query: When $\mathcal{A}$ sends a send query (i, c) to KEX, $\mathcal{A}_0$ sends a send query $(i, \mathcal{D}(c))$ to KEX_0. If $\mathcal{A}_0$ receives (i, m) it passes $(i, \mathcal{E}(c))$ along to $\mathcal{A}$.

Test query: If $\mathcal{A}$ sends test query to KEX, the constructed adversary $\mathcal{A}_0$ will use this as its challenge as well, passing the query to KEX_0. When $\mathcal{A}_0$ receives challenge (i, s_b) it forwards the challenge to $\mathcal{A}$.

Session key reveal: If $\mathcal{A}$ sends session reveal query to KEX, the query is passed to KEX_0. When $\mathcal{A}_0$ receives key (i, s_0) it forwards it to $\mathcal{A}$.

State reveal: If $\mathcal{A}$ sends state reveal query to KEX, the query is forwarded to KEX_0. When $\mathcal{A}_0$ receives random tape from KEX_0, it forwards this to $\mathcal{A}$.

Long term key reveal: If $\mathcal{A}$ sends long term reveal query to KEX, $\mathcal{A}_0$ passes query along to KEX_0. When $\mathcal{A}_0$ receives key pair $(\mathsf{pk}, \mathsf{sk})$, the key pair is forwarded to $\mathcal{A}$.

Since the challenges for $\mathcal{A}_0$ is the same as for $\mathcal{A}$, and the noise will not affect the session key value, it is clear that if $\mathcal{A}$ outputs the correct b then so does $\mathcal{A}_0$. Furthermore, if a session in $\mathsf{Exp}^{\mathsf{ROR}}_{\mathrm{KEX}}(\mathcal{A})$ is fresh then it also fresh in $\mathsf{Exp}^{\mathsf{ROR}}_{\mathrm{KEX}_0}(\mathcal{A}_0)$ since the sent and received messages all decode to the same values. Similarly, if $\mathcal{A}$ breaks implicit authentication if follows directly by construction that implicit authentication is broken in $\mathsf{Exp}^{\mathsf{ROR}}_{\mathrm{KEX}_0}(\mathcal{A}_0)$.

Combining the two reductions we get

$$\mathsf{Adv}^{\mathsf{ror}}_{1-\mathrm{KEX}}(\mathcal{A}) = \mathsf{Adv}^{\mathsf{ror}}_{\mathrm{KEX}_0}(\mathcal{A}_0).$$

□

5 Tools for Constructing Noisy Key Exchange

Now that we know we have a valid model we desire to develop the tools often used when designing key exchange protocols. The end goal is to develop a KEA [9] like construction that transforms an error-resistant asymmetric scheme through a modified FO-transform [6] into an error-resistant key encapsulation (KEM). The KEM is then used to construct an error-resistant key exchange protocol.

5.1 Other Error-Resistant Security Notions

The first step to achieve this is to define what we mean with an error-resistant asymmetric scheme and likewise an error-resistant KEM.

When working with error-resistant key exchange protocols we wanted a scheme that would remain functional even when transmitted messages contain noise. For asymmetric encryption schemes and KEMs this translates to schemes that should produce the correct output even when ciphertext contains some small amount of noise.

Definition 5. *Let $\Pi = (\mathfrak{K}, \mathcal{E}, \mathcal{D})$ be an asymmetric encryption scheme (key encapsulation mechanism) and let d be a metric. We say that Π is e-*error-resistant *over d if any two ciphertexts c_1, c_2 generated under the same public key pair $(\mathsf{sk}, \mathsf{pk})$ decrypt (decapsulate) to the same message m (key k) when $d(c_1, c_2) \leq e$.*

The security notions that revolve around revealing plain text will still hold in an unreliable setting. However, all security notions that allow ciphertexts to be revealed must be altered in order to prevent trivial attacks from the adversary. What we need is a modified IND-CCA notion that allows an adversary to reveal ciphertexts as long as they are not to close to the challenged ciphertext.

Since the KEA construction only requires that the KEM is IND-CCA secure, the following definitions will be defined for KEMs. A very similar definition can be made for asymmetric schemes as well.

Definition 6. *Let $\Pi = (\mathfrak{K}, \mathcal{E}, \mathcal{D})$ be a e-error-resistant key encapsulation mechanism (KEM) over d, let $b \leftarrow\!\!\$ \{0,1\}$, and let $\mathcal{A}$ be an adversary against Π that has access to encapsulation oracle $\mathcal{O}_\mathcal{E}$ and decapsulation oracle $\mathcal{O}_\mathcal{D}$. We define $\mathsf{Exp}_{e-\Pi}^{IND\text{-}bCCA}$ as follows:*

$\mathsf{Exp}_{e-\Pi}^{IND\text{-}bCCA}$	$\mathcal{O}_\mathcal{E}(\mathsf{pk}' \neq \mathsf{pk})$
1 : $b \leftarrow\!\!\$ \{0,1\}$	1 : **return** $(c', \mathsf{k}') \leftarrow \mathcal{E}(\mathsf{pk})$
2 : $\mathsf{k}_1 \leftarrow\!\!\$ \mathfrak{K}$	
3 : $(\mathsf{sk}, \mathsf{pk}) \leftarrow \mathfrak{K}$	$\mathcal{O}_\mathcal{D}(c')$, *where* $d(c, c') > e$
4 : $(c, \mathsf{k}_0) \leftarrow \mathcal{E}(\mathsf{pk})$	1 : **return** $\mathsf{k}' \leftarrow \mathcal{D}(\mathsf{sk}, c')$
5 : $b' \leftarrow \mathcal{A}^{\mathcal{O}_\mathcal{D}, \mathcal{O}_\mathcal{E}}(\mathsf{pk}, c, \mathsf{k}_b)$	

The adversary interacts with $\mathsf{Exp}_{e-\Pi}^{IND\text{-}bCCA}$, the encapsulation and decapsulation oracles and outputs a guess $b' \in \{0,1\}$ when it terminates. $\mathcal{A}$ may not query the decryption oracle on any ciphertext c' where $d(c, c') \leq e$. Define

$$\mathsf{Adv}_{e-\Pi}^{ind\text{-}bcca}(\mathcal{A}) = 2\left|\Pr[\mathcal{A} = b] - \frac{1}{2}\right|.$$

Π is said to be e-benign IND-CCA secure if $\mathsf{Adv}_{e-\Pi}^{ind\text{-}bcca}(\mathcal{A})$ is negligible for any polynomial time adversary $\mathcal{A}$.

Instead of determining if the challenge key received is the correct key, another way that results in breaking the key encapsulation scheme is to find two ciphertexts that generate the same session key but are further than e apart. It is clear that if two such ciphertexts are found for a specific key, then when the KEM is used as part of an error-resistant key exchange, one can be challenged while the other one queried.

Definition 7. *Let $\Pi = (\mathfrak{K}, \mathcal{E}, \mathcal{D})$ be an e-error-resistant key encapsulation mechanism (KEM) over d, and let $\mathcal{A}$ be an adversary against Π. Define $\mathsf{Exp}_{\Pi}^{\alpha-dist}$ as follows:*

$\mathsf{Exp}_{\Pi}^{\alpha-dist}$	
1 :	$(\mathsf{sk}, \mathsf{pk}) \leftarrow \mathfrak{K}$
2 :	$(c, \mathsf{k}) \leftarrow \mathcal{E}(\mathsf{pk})$
3 :	$(c', c'') \leftarrow \mathcal{A}(\mathsf{pk}, \mathsf{k})$

The adversary interacts with $\mathsf{Exp}_{\Pi}^{\alpha-dist}$ and outputs a guess (c', c'') when it terminates. Let E be the event that $d(c', c'') > \alpha$ and both c' and c'' produce k under valid key pairs $(\mathsf{sk}, \mathsf{pk})$ and $(\mathsf{sk}', \mathsf{pk})$. We say that $\mathcal{A}$ wins if E occurs and define

$$\mathsf{Adv}_{e-\Pi}^{\alpha-dist}(\mathcal{A}) = \Pr[\mathsf{E}].$$

The advantage of the adversary does of course not mean much in terms of protocol security if α is chosen to be less than e.

It is also worth pointing out that the secret key used in decryption is never specified. As long as the adversary finds two distinct ciphertexts c, c' that decrypts to c then the decryptions may use different secret keys $\mathsf{sk}, \mathsf{sk}'$ as long as they have the same public keys. The reason for this seemingly strange concept is that we wish to capture the chance of correct decryption given publicly available information.

5.2 Error Tolerant FO-Transform

The Fujisaki-Okamoto-transform is one of the most useful tools for developing key exchange protocols. It takes an asymmetric scheme with low security and transforms it through two steps into a IND-CCA secure KEM.

A vital part of the FO-transform, and part of why the security increases, is that it, as part of decryption, preforms a re-encryption in order to check that the received ciphertext is the same as the ciphertext produced by the decrypted plain text. If this is not the case the decryption algorithm will return $\perp$.

It is clear that this test will fail when used over error-resistant asymmetric schemes. In order to benefit from the FO-transform we need offer some modifications where cipher texts within a certain distance from each to other are allowed.

Definition 8 (Modified FO-transform). *The modified FO-transform takes a probabilistic e-error-resistant PKE-scheme $\Pi_0 = (\mathfrak{K}, \mathcal{E}_0, \mathcal{D}_0)$ and transforms it to a deterministic e-error-resistant PKE-scheme $\Pi_1 = (\mathfrak{K}, \mathcal{E}_1, \mathcal{D}_1)$ using the following encryption and decryption algorithms:*

$$\mathcal{E}_1(\mathsf{pk}, m) = \mathcal{E}_0(\mathsf{pk}, m; G(m))$$

$$\mathcal{D}_1(\mathsf{sk}, c) = \begin{cases} m' \leftarrow \mathcal{D}_0(\mathsf{sk}, c) \\ c' \leftarrow \mathcal{E}_0(\mathsf{pk}, m'; G(m')) \\ \\ \textbf{return } m', & \text{if } d(c', c) \leq e \\ \textbf{return } \perp, & \text{otherwise} \end{cases}$$

Then using the following approach a deterministic e-error-resistant PKE-scheme $\Pi_1 = (\mathfrak{K}, \mathcal{E}_1, \mathcal{D}_1)$ can be transformed into an e-error-resistant KEM $\Pi = (\mathfrak{K}, \mathcal{E}, \mathcal{D})$.

$$\mathcal{E}(\mathsf{pk}) = (c \leftarrow \mathcal{E}_1(\mathsf{pk}, m),\ \mathsf{k} \leftarrow H(c, m))$$

$$\mathcal{D}(\mathsf{sk}, c) = \begin{cases} H(\mathcal{E}_1(\mathsf{pk}, m'), m') & \text{if } m' \leftarrow \mathcal{D}_1(\mathsf{sk}, c) \\ \perp & \text{if } \perp \leftarrow \mathcal{D}_1(\mathsf{sk}, c) \end{cases}$$

where m is selected at random and H, G are hash functions in the random oracle model.

To see that this modified FO-transform has the desired effect we show that the error-resistance is preserved through each evolution step in the transformation. The proof of the following theorem closely follows the original FO-transform proof, the modifications can be found in Appendix A.

Theorem 2. *If $(\mathfrak{K}, \mathcal{E}_0, \mathcal{D}_0)$ is e-resistant and OW-CPA secure then $(\mathfrak{K}, \mathcal{E}_1, \mathcal{D}_1)$ created through the modified FO-transform is e-resistant and OW-PCA secure and $(\mathfrak{K}, \mathcal{E}, \mathcal{D})$ is e-benign IND-CCA secure.*

To see if the modified FO-transform has the desired effect we additionally need to determine how easy an adversary may find ciphertexts with colliding keys for the constructed KEM.

The following lemma indicates the probability that an adversary $\mathcal{C}$ can find two ciphertexts that both decrypt to the same session key independent of the secret key associated with the public key. It turns out this has no higher success rate than the adversary performing a lucky guess.

Lemma 1. *Let $\Pi = (\mathfrak{K}, \mathcal{E}, \mathcal{D})$ be an e-error-resistant KEM produced from an e-error-resistant PKE $\Pi_0 = (\mathfrak{K}, \mathcal{E}_0, \mathcal{D}_0)$ using the modified FO-transform. Let $\mathcal{C}$ be an adversary against Π. We then have*

$$\mathsf{Adv}^{2e-dist}_{e-\Pi}(\mathcal{C}) \leq \frac{l_H^2}{|\mathfrak{K}|},$$

where l_H denotes the number of hash-queries made by $\mathcal{C}$.

Proof. Assume there exists two secret keys $\mathsf{sk}, \mathsf{sk}'$ that have the same public key pk and produces the same key k under $\mathcal{E}$. Then

$$\mathcal{D}(\mathsf{sk}, c) = \mathcal{D}(\mathsf{sk}', c').$$

This means

$$H(\mathcal{E}_1(\mathsf{pk}, m), m) = H(\mathcal{E}_1(\mathsf{pk}, m'), m') \tag{1}$$

where $m = \mathcal{D}_1(\mathsf{sk}, c)$ and $m' = \mathcal{D}_1(\mathsf{sk}', c')$. We know that if 1 holds then either $m = m'$ or there is a collision in H.

If the first case is true this means

$$d(c, \mathcal{E}_0(\mathsf{pk}, m; G(m))) \leq e \qquad d(c', \mathcal{E}_0(\mathsf{pk}, m; G(m))) \leq e,$$

where it from the triangle inequality property of metrics directly follows that

$$d(c, c') \leq 2e,$$

which is still smaller the minimum distance required in order for $\mathsf{Exp}^{2e-\text{dist}}$ to be satisfied.

This means the only chance the adversary has of success is to find a collision on H. Due to the birthday bound the advantage of the adversary is bounded by $\mathsf{Adv}^{2e-\text{dist}}_{\Pi-e}(\mathcal{C}) \leq \frac{l_H^2}{|\mathfrak{K}|}$. □

5.3 Error-Resistant KEA Construction

We now have both the language and the tools needed to finally build the error-resistant key exchange from an error-resistant KEM.

Theorem 3. *Let $\Pi = (\mathfrak{K}, \mathcal{E}, \mathcal{D})$ be an error-resistant KEM generated from an e-error-resistant PKE over d, i.e. Π_0, through the modified FO-transform.*

Then the key exchange protocol, KEX*, with following initiator and responder algorithms is an e-error-resistant key exchange protocol over d.*

$\mathcal{I}$	$\mathcal{R}$
1 : $(c_{\mathcal{R}}, \mathsf{k}_{\mathcal{R}}) \leftarrow \mathcal{E}(\mathsf{pk}_{\mathcal{R}})$	1 : *Await until* $c'_{\mathcal{R}}$ *received from* $\mathcal{I}$
2 : *Send* $c_{\mathcal{R}}$ *to* $\mathcal{R}$	2 : $(c_{\mathcal{I}}, \mathsf{k}_{\mathcal{I}}) \leftarrow \mathcal{E}(\mathsf{pk}_{\mathcal{I}})$
3 : *Await until* $c'_{\mathcal{I}}$ *received from* $\mathcal{R}$	3 : *Send* $c_{\mathcal{I}}$ *to* $\mathcal{I}$
4 : $\mathsf{k}'_{\mathcal{I}} \leftarrow \mathcal{D}(\mathsf{sk}_{\mathcal{I}}, c'_{\mathcal{I}})$	4 : $\mathsf{k}'_{\mathcal{R}} \leftarrow \mathcal{D}(\mathsf{sk}_{\mathcal{R}}, c'_{\mathcal{R}})$
5 : $\mathsf{k} \leftarrow kdf(\mathsf{k}'_{\mathcal{I}}, \mathsf{k}_{\mathcal{R}}, \mathsf{pk}_{\mathcal{R}}, \mathsf{pk}_{\mathcal{I}})$	5 : $\mathsf{k} \leftarrow kdf(\mathsf{k}_{\mathcal{I}}, \mathsf{k}'_{\mathcal{R}}, \mathsf{pk}_{\mathcal{R}}, \mathsf{pk}_{\mathcal{I}})$

Furthermore, if $\mathcal{A}$ is an adversary against KEX *that can perform no state queries, l_h hash queries, l_p key generation queries, l_i execute queries and 1 test query, then there exists an adversary $\mathcal{B}$ against Π such that*

$$\mathsf{Adv}^{\mathsf{ror}}_{e-\mathrm{KEX}}(\mathcal{A}) \leq l_i l_p \mathsf{Adv}^{ind\text{-}bcca}_{e-\Pi}(\mathcal{B}) + \frac{l_h^2}{|\mathfrak{K}|}.$$

Proof (Proof Sketch).

In the first scenario the adversary sends a session reveal query to an instance that has queried *kdf* on the correct input, i.e. $\mathcal{A}$ sends a session reveal query to an instance with knowledge of the challenged session key. Since the public keys are part of *kdf* used to compute the session keys, the only viable instances to query are those instances that agree on the public keys used during negotiation. From Lemma 1 we know that given an e-error-resistant KEM the probability that an adversary may find two ciphertexts at least than $2e$-distance apart is upper bound by $\frac{l_H^2}{|\mathfrak{K}|}$. Since we have good distance bound e we know that any instances with transcripts closer than $2e$ are loosely related. As such we have that the adversarial advantage against scenario one is upper bounded by $\frac{l_H^2}{|\mathfrak{K}|}$, otherwise freshness does not hold.

In the second scenario the adversary will have to determine the KEM-generated keys. There are two ways to determine the KEM-keys k_i; either by breaking Π, or through sufficient queries. Gaining knowledge of both KEM-keys through queries alone is not possible without breaking freshness. This leaves breaking the underlying KEM protocol.

A reduction from $\mathcal{A}$ to a KEM-adversary $\mathcal{B}$ is easily achieved as follows

$\mathcal{B}$ **on challenge** $(\mathsf{pk}, c, \mathsf{k})$: when $\mathcal{A}$ returns b, $\mathcal{B}$ returns b.
  Select an instance j and user j at random.
  Run $\mathcal{A}$ and simulate KEX according to protocol with the following alterations
    if $\mathcal{A}$ sends a key generation query for user i:
    Send pk to $\mathcal{A}$
    if $\mathcal{A}$ sends execute query for instance j:
    Send (j, c) to $\mathcal{A}$
    if $\mathcal{A}$ sends test query to instance s:
      if the role of s is $\mathcal{I}$ and ($s = j$ or s, j are partners):
        Send $kdf(\mathsf{k}, \mathsf{k}_{\mathcal{R}}, \mathsf{pk}_s, \mathsf{pk}_{\mathcal{R}})$
      if the role of s is $\mathcal{R}$ and ($s = j$ or s, j are partners):
        Send $kdf(\mathsf{k}_{\mathcal{I}}, \mathsf{k}, \mathsf{pk}_{\mathcal{I}}, \mathsf{pk}_s)$

If $\mathcal{B}$ selects the correct user and instance that $\mathcal{A}$ will query then a correct answer from $\mathcal{A}$ will yield a correct answer for $\mathcal{B}$. If the isntances do not match the success probability of $\mathcal{B}$ is $\frac{1}{2}$. Since the user and instance are chosen completely at random without adversary knowledge, the probability that one of the two cases above happens is $\frac{1}{l_u l_s}$. We therefore get that

$$Adv(\mathcal{A}) \leq l_u l_s Adv(\mathcal{B}) + \frac{l_h^2}{|\mathfrak{K}|}.$$

□

6 Existence of Error-Resistant PKE

Finally, we need to show the existence of e-error-resistant public key cryptosystems. As it turns out, McEliece is an OW-CPA PKE that uses Goppa Codes. By tweaking the parameters we are able to use the integrated error correcting codes both to correct against errors as well as still preserve the OW-CPA functionality.

$\mathfrak{K}(1^t, 0^e)$

Generate the three matrices G, S, P.
G a $k \times n$ Goppacode that can correct ut to t errors.
$\psi()$ an efficient decoding algorithm of G.
S a $k \times k$ binary, nonsignular random matrix.
P a $n \times n$ permutation matrix.
$\mathsf{sk} \leftarrow (S, G, P, \psi())$
$\mathsf{pk} \leftarrow (SGP, t, e)$

$\mathcal{E}(\mathsf{pk}, m)$

$z \leftarrow \mathcal{B}^n$ // Hamiltonian weight of z equals $t - e + 1$
$c \leftarrow m \cdot \mathsf{pk} \oplus z = mSGP \oplus z$

$\mathcal{D}(\mathsf{sk}, c)$

$mS \leftarrow \psi(cP^{-1}) = \psi((mSGP \oplus z \oplus \epsilon)P^{-1}) = \psi(mSG \oplus (z \oplus \epsilon)P^{-1})$
$m' \leftarrow mS(S^{-1})$

Fig. 2. OW-CPA secure McEliece [8] modified to obtain e-resistance.

Theorem 4. *The modified McEliece protocol presented in Fig. 2 is OW-CPA secure and e-resistant using hamming distance metric.*

Proof. To see that the protocol is e-error-resistant let $(\mathsf{sk} = (S, G, P, \psi()), \mathsf{pk} = (SGP, t, e))$ be an arbitrary key pair. Then for any message m

$$c = \mathcal{E}(pk, m) = mSGP \oplus z.$$

Let $\epsilon \in \mathcal{B}^n$ be a bit-string of length n with hamming weight at most e be an arbitrary error vector. Decrypting we get

$$m' = \mathcal{D}(sk, c + \epsilon) = \psi(mSG \oplus (z \oplus \epsilon)P^{-1})S^{-1},$$

and since $z \oplus \epsilon$ has Hamming weight at most t and P is a permutation matrix, we have that ψ corrects $z \oplus \epsilon$. Thus we get

$$m' = mSS^{-1} = m$$

and we have that the modified McEliece from Fig. 2 is e-error-resistant.

For large enough values of n, t, k and small enough e the original proof of OW-CPA security holds [8,13]. □

We now know that an error-resistant PKE does exist and our proposed protocol does indeed work in theory.

As final remarks we would like to comment on the fact that the final protocol presented is not a practical solution. Since we cannot assume that all public keys are stored locally on each device these keys would need to be transmitted and versified prior to use. The length of said public keys are, however, too large to be reliably transmitted over unreliable networks [1]. Combining the substantial key length with the short lifespan of emergency network messages, it would be more efficient to use a Diffie-Hellman based approach that requires perfect transmission in these settings. This is, however, an approach that is currently considered unfeasible. In other words; for KEX to be applicable over unreliable networks a great deal of work is needed in order to reduce message and key size.

A Proof of Modified FO-transform Theorem 2

The proof sketch below has to be read in correspondence with the original proof in Hofheinz, Hövelmanns and Kiltz [6]. The only alterations needed, and thus given, are those of each game-step. The remaining arguments hold with respect to the new games presented.

Proof (Proof sketch).
PKE OW-CPA $\implies$ PKE$_1$ OW-PCA:
The original reduction and proof given by Hofheinz, Hövelmanns and Kiltz [6] holds for the new game setup presented below when using e-error-resistant PKEs

for correctness. The only difference between the original setup and the modified setup is that we allow e amount of noise on every ciphertext instead of requiring strict equality (Fig. 3).

PKE_1 OW-PCA $\implies$ $\text{KEM}^{\perp}$ IND-CCA:
The hash oracle for H requires a larger rewrite since we need to make sure the oracle is consistent when given ciphertext less than e apart. Other than some bookkeeping this does not infringe on the argument and the original proof by Hofheinz, Hövelmanns and Kiltz [6] still holds (Fig. 4).

Games $G_0 - G_3$	$e - PCO(m, c)$
$(\mathsf{sk}, \mathsf{pk}) \leftarrow \mathcal{K}$	$m' \leftarrow \mathcal{D}(\mathsf{sk}, c)$ // G0-G1
$m^* \leftarrow\$ \mathcal{M}$	**return** $([m = m']$ **and** // G0-G1
$c^* \leftarrow \mathcal{E}(\mathsf{pk}, m^*)$	$[d(c, \mathcal{E}(\mathsf{pk}, m'; G(m'))) < e])$
$m' \leftarrow \mathcal{B}(\mathsf{pk}, c^*)$	**return** $[d(c, \mathcal{E}(\mathsf{pk}, m; G(m))) < e]$
return $[m' = m^*]$	

G(m)	$e - CVO(c)$, where $d(c, c^*) \geq e$
if $\exists r((m, r) \in \mathfrak{L}_G)$	$m' \leftarrow \mathcal{D}(\mathsf{sk}, c)$ // G0-G1
return r	**return** $([m' \in \mathcal{M}]$ **and** // G0
if $m = m^*$ // G3	$[d(c, \mathcal{E}(\mathsf{pk}, m'; G(m'))) < e])$
$QUERY :=$ **true** // G3	**return** $(\exists (m, r) \in \mathfrak{L}_G$ **and** // G1
abort // G3	$[d(c, \mathcal{E}(\mathsf{pk}, m; r)) < e]$ **and**
$r \leftarrow\$ \mathcal{R}$	$[m' = m])$
$\mathfrak{L}_G = \mathfrak{L}_G \cup \{(m, r)\}$	**return** $(\exists (m, r) \in \mathfrak{L}_G$ **and** // G2-G3
return r	$[d(c, \mathcal{E}(\mathsf{pk}, m; r)) < e])$

Fig. 3. PKE OW-CPA $\implies$ PKE_1 OW-PCA. Where $\mathcal{B}$ is an adversary of the form $B^{G(\cdot), PCO(\cdot,\cdot), CVO(\cdot)}$

Games $G_0 - G_2$

$(\mathsf{sk}, \mathsf{pk}) \leftarrow \mathcal{K}$
$m^* \leftarrow\$ \mathcal{M}$
$c^* \leftarrow \mathcal{E}_1(\mathsf{pk}, m^*)$
$\mathsf{k}_0^* := H(m^*, c^*)$
$\mathsf{k}_1^* \leftarrow\$ \mathfrak{K}$
$b \leftarrow\$ \{0,1\}$
$b' \leftarrow B^{\mathcal{D}^{\perp}, H}(\mathsf{pk}, c^*, k_b^*)$
return $[b' = b]$

$\mathcal{D}^{\perp}(c)$, where $d(c, c^*) \geq e$ // G0

$m' := \mathcal{D}_1(\mathsf{sk}, c)$
if $m' = \perp$ **return** $\perp$
return $\mathsf{k} := H(m', c)$

$\mathcal{D}^{\perp}(c)$, where $d(c, c^*) \geq e$ // G1-G2

if $\exists \mathsf{k}$ s.t $(c, \mathsf{k}) \in \mathfrak{L}_D$ **return** k
if $\mathcal{D}_1(\mathsf{sk}, c) \notin \mathcal{M}$ **return** $\perp$
$\mathsf{k} \leftarrow\$ \mathcal{K}$
$\mathfrak{L}_D := \mathfrak{L}_D \cup \{(c, \mathsf{k})\}$
return k

$H(m, c)$

if $\mathcal{D}_1(\mathsf{sk}, c) = \perp$
  return $\perp$
if $\exists \mathsf{k}$, s.t. $(m, \mathcal{E}_1(\mathsf{pk}, m), \mathsf{k}) \in \mathfrak{L}_H$
  return k
$\mathsf{k} \leftarrow\$ \mathcal{K}$
if $d(c, c^*) < e$ // G2
  $CHAL :=$ **true**
  abort
if $\exists \mathsf{k}'(\mathcal{E}_1(\mathsf{pk}, m), \mathsf{k}') \in \mathfrak{L}_D$ // G1-G2
  $\mathsf{k} := \mathsf{k}'$
else // G1-G2
  $\mathfrak{L}_D := \mathfrak{L}_D \cup \{(\mathcal{E}_1(\mathsf{pk}, m), \mathsf{k})\}$
$\mathfrak{L}_H := \mathfrak{L}_H \cup \{(m, \mathcal{E}_1(\mathsf{pk}, m), k)\}$
return k

Fig. 4. PKE_1 OW-PCA $\implies$ $\mathrm{KEM}^{\perp}$ IND-CCA

References

1. Albrecht, M.R., et al.: Classic McEliece: conservative code-based cryptography. https://classic.mceliece.org/nist/mceliece-20201010.pdf
2. Bellare, M., Rogaway, P.: Entity authentication and key distribution. In: Stinson, D.R. (ed.) CRYPTO 1993. LNCS, vol. 773, pp. 232–249. Springer, Heidelberg (1994). https://doi.org/10.1007/3-540-48329-2_21
3. Day, J., Zimmermann, H.: The OSI reference model. Proc. IEEE **71**(12), 1334–1340 (1983). https://doi.org/10.1109/PROC.1983.12775
4. Eddy, W.: Transmission Control Protocol (TCP). RFC 9293, August 2022. https://doi.org/10.17487/RFC9293, https://www.rfc-editor.org/info/rfc9293
5. Gligoroski, D., Knapskog, S.J., Andova, S.: Cryptcoding - encryption and error-correction coding in a single step. In: Security and Management (2006)
6. Hofheinz, D., Hövelmanns, K., Kiltz, E.: A modular analysis of the Fujisaki-Okamoto transformation. In: Kalai, Y., Reyzin, L. (eds.) TCC 2017. LNCS, vol. 10677, pp. 341–371. Springer, Cham (2017). https://doi.org/10.1007/978-3-319-70500-2_12
7. Huffman, W.C., Pless, V.: Fundamentals of Error-Correcting Codes. Cambridge University Press, Cambridge (2003)

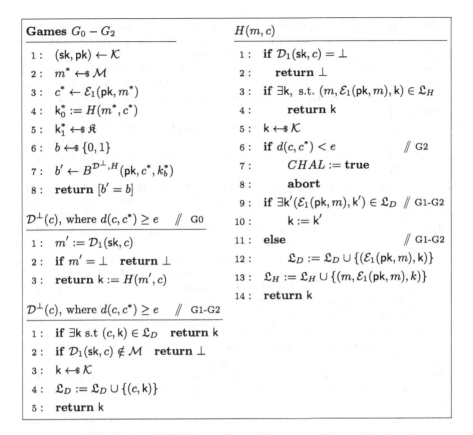

8. Kobara, K., Imai, H.: On the one-wayness against chosen-plaintext attacks of the Loidreau's modified McEliece PKC. IEEE Trans. Inf. Theory **49**(12), 3160–3168 (2003). https://doi.org/10.1109/TIT.2003.820016
9. Lauter, K., Mityagin, A.: Security analysis of KEA authenticated key exchange protocol. In: Yung, M., Dodis, Y., Kiayias, A., Malkin, T. (eds.) PKC 2006. LNCS, vol. 3958, pp. 378–394. Springer, Heidelberg (2006). https://doi.org/10.1007/11745853_25
10. Li, B., Micciancio, D.: On the security of homomorphic encryption on approximate numbers. In: Canteaut, A., Standaert, F.-X. (eds.) EUROCRYPT 2021. LNCS, vol. 12696, pp. 648–677. Springer, Cham (2021). https://doi.org/10.1007/978-3-030-77870-5_23
11. Li, Y., Schäge, S.: No-match attacks and robust partnering definitions: defining trivial attacks for security protocols is not trivial. In: Proceedings of the 2017 ACM SIGSAC Conference on Computer and Communications Security, CCS 2017, pp. 1343–1360. Association for Computing Machinery, New York, NY, USA (2017). https://doi.org/10.1145/3133956.3134006
12. Lien, Y.N., Jang, H.C., Tsai, T.C.: A MANET based emergency communication and information system for catastrophic natural disasters. In: 2009 29th IEEE International Conference on Distributed Computing Systems Workshops, pp. 412–417 (2009). https://doi.org/10.1109/ICDCSW.2009.72
13. McEliece, R.J.: A public-key cryptosystem based on algebraic coding theory. Deep Space Network Progress Report, pp. 114–116 (1978)
14. Postel, J.: User Datagram Protocol. RFC 768, August 1980. https://doi.org/10.17487/RFC0768, https://www.rfc-editor.org/info/rfc768
15. Quispe, L.E., Mengual, L.: Behavior of Ad Hoc routing protocols, analyzed for emergency and rescue scenarios, on a real urban area. Expert Syst. Appl. **41**, 2565–2573 (2014)
16. Rao, T.R.N.: Joint encryption and error correction schemes. SIGARCH Comput. Archit. News **12**(3), 240–241 (1984). https://doi.org/10.1145/773453.808188
17. TCCA: Voice & data. https://tcca.info/tetra/tetra-your-service/voice-data/

Generic Error SDP and Generic Error CVE

Felice Manganiello and Freeman Slaughter(✉)

Clemson University, Clemson, SC, USA
{manganm,fslaugh}@clemson.edu

Abstract. This paper introduces a new family of CVE schemes built from generic errors (GE-CVE) and identifies a vulnerability therein. To introduce the problem, we generalize the concept of error sets beyond those defined by a metric, and use the set-theoretic difference operator to characterize when these error sets are detectable or correctable by codes. We prove the existence of a general, metric-less form of the Gilbert-Varshamov bound, and show that - like in the Hamming setting - a random code corrects a generic error set with overwhelming probability. We define the generic error SDP (GE-SDP), which is contained in the complexity class of NP-hard problems, and use its hardness to demonstrate the security of GE-CVE. We prove that these schemes are complete, sound, and zero-knowledge. Finally, we identify a vulnerability of the GE-SDP for codes defined over large extension fields and without a very high rate. We show that certain GE-CVE parameters suffer from this vulnerability, notably the restricted CVE scheme.

Keywords: Code-based cryptography · Syndrome Decoding Problem · generic error set · zero-knowledge scheme · CVE

1 Introduction

In 2016, NIST - recognizing the lack of acceptably secure post-quantum cryptographic protocols - created a competition of sorts, comparing the pros and cons of proposed algorithms, with the aim on creating a system that is secure against both classical and quantum computers. Code-based cryptography has become an attractive post-quantum candidate over the years, as it is believed to present a computationally difficult problem even against quantum computers [14]. Classic McEliece [3], a code-based KEM has recently passed into Round 4 of the NIST Post-Quantum Standardization Competition and remains the candidate based on the oldest problem in the competition [1].

The Cayrel-Véron-El Yousfi Alaoui (CVE) protocol is a zero-knowledge identification scheme based on the Syndrome Decoding Problem (SDP) for linear codes, with competitive computation speed and key sizes in practical instances. This is a fruitful system; it is quite flexible in that it can be applied to a range

Partially supported by the National Science Foundation under grant DMS-1547399.

A. Esser and P. Santini (Eds.): CBCrypto 2023, LNCS 14311, pp. 125–143, 2023.
https://doi.org/10.1007/978-3-031-46495-9_7

of different cryptographic primitives and settings. This paper focuses on generic error sets and the theory of error correctability and detectability based on the set difference operator contained in [28,29]. Then, to elucidate, we apply this work to two specific cases: the restricted form of CVE, which takes a different error set than the standard protocol, and rank-CVE, which takes a different metric. Using this framework, we generalize the SDP and CVE to generic error sets independent of metric, introducing a new, NP-complete SDP based on these arbitrary errors. From this, we can construct a generic error CVE. We characterize the parameters of error sets for which this generic SDP has a polynomial-time decoding algorithm, leading to a vulnerability of generic error CVE for certain error sets. To be clear, we show that CVE based on the restricted SDP is vulnerable when it is defined for certain parameters of codes without a high rate.

The paper is organized as follows. In Sect. 2, we recall standard coding theory results and present the notation we will use throughout this paper. Section 3 introduces the set difference, an operator from set theory. When this operator repeatedly acts on a set of generic errors, we show that it will stabilize at a subspace over a prime field. We introduce the notion of detectability and correctability using the set difference, resulting in a more general concept than the standard definition based on balls. These concepts also result in a generalization of the Gilbert-Varshamov Bound that guarantees the existence of a code correcting an arbitrary error set. Finally, we use the results of this section to prove that a random code will correct a general error set with a probability that tends towards one as the code length increases. Section 4 is devoted to complexity. From the SDP, we define the Generic Error SDP (GE-SDP) and its decisional variant and provide evidence that they are NP-complete problems. For Sect. 5, we generalize the zero-knowledge CVE scheme from [15] to generic errors and prove that it is complete, sound, and zero-knowledge. Paired with the complexity arguments of the previous section, we obtain strict bounds about the probability that an adversary can forge their veracity to a verifier in this generic error setting. Finally, we devote Sect. 6 to highlighting a vulnerability that results from this generic definition of detectability and correctability. We show that the GE-SDP can be solved in polynomial times under certain conditions. We apply this result to the Restricted Syndrome Decoding Problem (R-SDP) defined in [7] and show that for certain parameter choices, an adversary can correct the errors introduced to obfuscate the plaintext in a polynomial-time decoding algorithm.

2 Preliminaries

We introduce the notation used throughout this paper and recall some standard linear algebra and coding theory results.

Let $\mathbb{F}_q$ denote the finite field with q elements, with $q = p^N$ a prime power and $\mathbb{F}_q^n$ be the set of n-length vectors over $\mathbb{F}_q$. For $E \subseteq \mathbb{F}_q^n$, let E^* be $E \setminus \{0\}$. For a set $E \subseteq \mathbb{F}_q^n$ with k elements, let $\langle E \rangle_{\mathbb{F}_p}$ be the span of E over $\mathbb{F}_p$:

$$\langle E \rangle_{\mathbb{F}_p} = \lambda_1 e_1 + \lambda_2 e_2 + ... + \lambda_k e_k \text{ for } \lambda_i \in \mathbb{F}_p, e_i \in E.$$

Let $\mathfrak{M}_n$ be the set of monomial transformations. A monomial transformation $\tau : \mathbb{F}_q^n \to \mathbb{F}_q^n$ is a map that acts on vectors by permuting their entries and scaling them by non-zero multiples. That is, there exists $v \in \mathbb{F}_q^n \setminus \{0\}$ and $\sigma \in S_n$, the symmetric group, such that

$$\tau(x) = (v_{\sigma(1)}x_{\sigma(1)}, v_{\sigma(2)}x_{\sigma(2)}, \ldots, v_{\sigma(n)}x_{\sigma(n)}) \text{ for all } x \in \mathbb{F}_q^n.$$

Recall that $GL_n(\mathbb{F}_q)$ is the set of $n \times n$ invertible matrices over the field $\mathbb{F}_q$, which forms a group under standard matrix multiplication. We define the stabilizer of a set $E \subseteq \mathbb{F}_q^n$, denoted $\mathfrak{S}_E$, as the set of invertible matrices that map E into E, meaning $\mathfrak{S}_E = \{M \in GL_n(\mathbb{F}_q) \mid eM \in E \text{ for all } e \in E\}$.

We continue with some basic definitions and results from coding theory, which may be found in [32].

Definition 1. *We say C is an $[n, k]$-linear code when C is a linear subspace of $\mathbb{F}_q^n$ over $\mathbb{F}_q$ of dimension k.*

We focus this work on linear codes, and we refer to them simply as codes. We also assume that C is a code over $\mathbb{F}_q$.

Definition 2. *For an $[n, k]$ code C, a generator matrix $G \in \mathbb{F}_q^{k \times n}$ is a full-rank matrix where the rows are comprised of a basis of C over $\mathbb{F}_q$. The parity-check matrix of a code C is a (full-rank) matrix $H \in \mathbb{F}_q^{(n-k)\times n}$ such that $C = \{x \in \mathbb{F}_q^n \mid xH^t = 0\}$.*

For $x \in \mathbb{F}_q^n$, we define the Hamming weight $\omega(x)$ to be the number of non-zero entries in the vector x. This gives rise to the Hamming distance between two vectors, $d(x, y)$, defined as the weight of their difference $\omega(x - y)$. The Hamming ball with radius r and center x is denoted $B_r(x)$, and is defined as the set of vectors that are distance less than or equal to r from x. While we will not focus on the Hamming metric in this paper, it is used by tradition in certain definitions that we will generalize in future sections.

3 Generic Error Sets

In this section, we generalize the concepts of decodability and correctability with generic error sets beyond any metric. To do that, we need to introduce the concept of set difference.

Definition 3. *For some $E \subseteq \mathbb{F}_q^n$, the set difference of E is $\Delta E = \{e_1 - e_2 \mid e_1, e_2 \in E\}$.*

Since the set E will represent an error set for the purposes of our work, we assume that $0 \in E \subseteq \mathbb{F}_q^n$.

The following examples demonstrate that the cardinality of the set difference depends on if the elements themselves are in an arithmetic progression. This is a result of the Cauchy-Davenport Theorem for restricted sumsets; see [26, Theorem 25] and [27, Theorem 3]. We only use this example to show that the set difference is not immediate from the set itself and that one must inspect every element in the worst case to check for arithmetic progression.

Example 1. Consider $A = \{1,2,3\} \subseteq \mathbb{F}_7$, where the elements of A are in an arithmetic progression modulo 7. We can calculate $\Delta A = \{0,1,2,5,6\}$, thus $|\Delta A| = 2|A| - 1$.

Example 2. On the other hand, the elements of $B = \{1,2,4\}$ are not in an arithmetic progression. This time, despite having the same cardinality as A in Example 1, we see that $\Delta B = \{0,1,2,3,4,5,6\}$, with $|\Delta B| > 2|B| - 1$.

We introduce the concept of Δ-closure of a set $E \subseteq \mathbb{F}_q^n$, meaning the smallest set that contains E and all the difference sets originating by it.

Theorem 1. *For a set $E \subseteq \mathbb{F}_q^n$, the chain $E \subseteq \Delta E \subseteq \Delta^2 E \subseteq ...$ stabilizes, meaning that there exists some $k \in \mathbb{N}$ such that $\Delta^k E = \Delta^{k+1} E$. In this case, $\Delta^k E = \langle E \rangle_{\mathbb{F}_p}$.*

Proof. The chain stabilizes because we work with a finite sets.

Let $x \in \Delta^r E$ and $y \in \Delta^s E$ for $r, s \in \mathbb{N}$ with $r \geq s$. Then $-y \in \Delta^{r+1} E$, so $x + y = x - (-y) \in \Delta^{r+2} E$. Now for $x \in \Delta^r E$ and $\alpha \in \mathbb{F}_p$, then $\alpha x \in \Delta^{r+\alpha} E$. From this, we can see that for $x, y \in \Delta^k E = \Delta^{k+1} E$, we have that $x + y \in \Delta^k E$ and $\alpha x \in \Delta^k E$. Thus $\Delta^k E$ is an $\mathbb{F}_p$-subspace, so $\Delta^k E \subseteq \langle E \rangle_{\mathbb{F}_p}$.

On the other hand, for all $x \in \Delta^r E$, we can write $x = x_{r-1} - y_{r-1}$ for $x_{r-1}, y_{r-1} \in \Delta^{r-1} E$, so each element is the difference of two elements one step down on the difference chain. Continuing this, $x = \sum_{n=1}^{|E|} \alpha_n x_n$ for $x_i \in E$ and $\alpha_i \in \mathbb{F}_p$. Thus $\langle E \rangle_{\mathbb{F}_p} \subseteq \Delta^k E$, as every element in $\Delta^k E$ can be decomposed into a linear combination of elements from E.

Definition 4. *For a set $E \subseteq \mathbb{F}_q^n$, the Δ-closure of E is $\overline{E}^{\Delta} = \lim_{k \to \infty} \Delta^k E$. We say that a set $E \subseteq \mathbb{F}_q^n$ is Δ-closed if $E = \overline{E}^{\Delta}$.*

The cardinality of a Δ-closed set is as follows.

Corollary 1. *For a set $E \subseteq \mathbb{F}_q^n$, the $|\overline{E}^{\Delta}| = p^m$ for some $m \in \mathbb{N}$.*

This is a corollary of Theorem 1 where we show that the stabilizing set $\overline{E}^{\Delta}$ is a subspace of $\mathbb{F}_q^n$ over $\mathbb{F}_p$.

Example 3. Since we will work on Δ-closed sets throughout this manuscript, we provide two examples that are going to be used.

- $\overline{B_r(0)}^{\Delta} = \mathbb{F}_q^n$ for any $r > 0$.
- If $E = \{0,1\}^n \subseteq \mathbb{F}_q^n$, then $\overline{E}^{\Delta} = \mathbb{F}_p^n$.

3.1 Error Detectability and Correctability

When considering communication over a q-ary symmetric channel, errors are additive. More precisely, if $\mathcal{C} \subseteq \mathbb{F}_q^n$ is an $[n,k]$ code of minimum distance d and $c \in \mathcal{C}$ is sent through the channel $c + e$ with $e \in \mathbb{F}_q^n$ is received. Considering a minimum distance decoder, meaning a map that returns the unique closest codeword to a received vector if it exists, then we can say that an error is detectable if $e \in B_{d-1}(0)$, and (uniquely) correctable if $e \in B_t(0)$ with $t = \lfloor \frac{d-1}{2} \rfloor$. These are well-known results from classical coding theory that can be reviewed in any coding theory textbook such as [32].

With another application in mind, similar definitions have been developed based on the rank distance, meaning the rank over $\mathbb{F}_p$ between two elements of $\mathbb{F}_q^n$. This metric is useful when communicating over a multicast network; for more information, see [31] and [33].

In [28] and [29], the author generalizes the concepts of error detectability and correctability to generic error sets. We recall some of the results in this section for matters of completeness.

Definition 5. *An error set $E \subseteq \mathbb{F}_q^n$ is detectable by some code $\mathcal{C} \subseteq \mathbb{F}_q^n$ if $E \cap \mathcal{C} = \{0\}$, or equivalently if $E^* \cap \mathcal{C} = \emptyset$. Similarly, this set of errors E is correctable by $\mathcal{C}$ if $\Delta E \cap \mathcal{C} = \{0\}$.*

This definition generalizes the classical concept of detectability and correctability based on Hamming balls and balls based on rank metric. In the case of Hamming balls,

$$\Delta B_t(0) \subseteq B_{d-1}(0), \tag{1}$$

meaning that any error that is detectable under the difference set definition is also detectable under the minimum distance of a code. Note that the difference set of a ball is a ball itself: if d is odd, then $\Delta B_t(0) = B_{d-1}(0)$, whereas if d is even, then $\Delta B_t(0) = B_{d-2}(0)$.

The following proposition can be viewed as a motivation for the language used in Definition 5.

Proposition 1. *Let $\mathcal{C} \subseteq \mathbb{F}_q^n$ be a code with parity-check matrix $H \in \mathbb{F}_q^{n-k \times n}$. The set $E \subseteq \mathbb{F}_q^n$ is correctable by $\mathcal{C}$ if and only if its syndromes are unique, meaning that for $e, e' \in E$, $eH^t = e'H^t$ if and only if $e = e'$.*

Proof. Let $e, e' \in E$ be errors with the same syndrome, meaning that $eH^t = e'H^t$. This is equivalent to saying that $(e - e')H^t = 0$, meaning that $e - e' \in \mathcal{C}$. Since by hypothesis $\Delta E \cap \mathcal{C} = \{0\}$, then $e = e'$.

The following proposition is a direct consequence of Definition 5.

Proposition 2. *Let $\mathcal{C} \subseteq \mathbb{F}_q^n$ be a code. $E \subseteq \mathbb{F}_q^n$ is decodable by $\mathcal{C}$ if and only if ΔE is detectable by $\mathcal{C}$.*

It follows the next corollary showing that Δ-closed sets are maximal sets for which detectability corresponds to correctability.

Corollary 2. *Given a code $\mathcal{C} \subseteq \mathbb{F}_q^n$, a set $E \subseteq \mathbb{F}_q^n$ is detectable and correctable if and only if E is Δ-closed, meaning that $\overline{E}^{\Delta} = E$.*

We focus now on results regarding the existence of codes correcting a generic error set $E \subseteq \mathbb{F}_q^n$.

3.2 Generic Gilbert-Varshamov Bound

The proof of the following results may be found in [28] or [29], and are based on the concept of a balanced family of codes.

Definition 6. *Let $\mathcal{B}$ be a collection of $[n, k]$-linear codes. We call $\mathcal{B}$ a balanced family of codes if each vector in $(\mathbb{F}_q^n)^*$ belongs to the same number of codes of $\mathcal{B}$.*

Theorem 2 ([28]). *Let $\mathcal{B}$ be a balanced family of codes and $f : \mathbb{F}_q^n \to \mathbb{C}$ be a complex-valued function. Then*

$$\frac{1}{|\mathcal{B}|}\sum_{\mathcal{C}\in\mathcal{B}}\sum_{c\in\mathcal{C}^*} f(c) = \frac{q^k-1}{q^n-1}\sum_{v\in(\mathbb{F}_q^n)^*} f(v).$$

Proof. Construct a bipartite graph where the upper nodes are linear codes in $\mathcal{B}$, the lower nodes are the non-zero elements of $\mathbb{F}_q^n$, and there is an edge if the code above contains the element below. Figure 1 depicts such a bipartite graph.

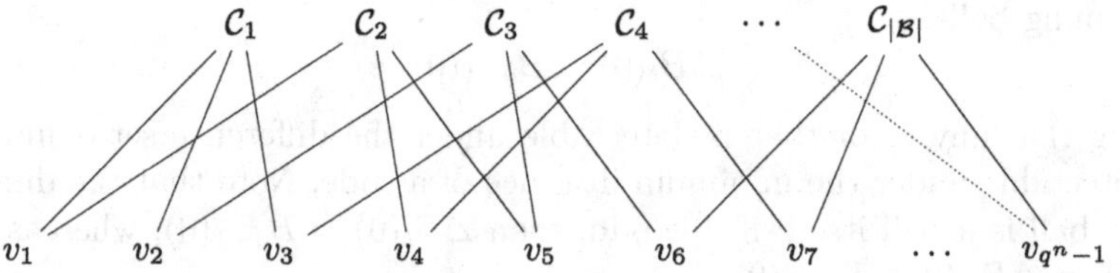

Fig. 1. Bipartite graph of a balanced family.

There are $|\mathcal{B}|$ nodes above and $q^n - 1$ nodes below. This graph is regular, meaning each top node has the same degree of $q^k - 1$, and each bottom node has the same degree $N_\mathcal{B}$. Counting the edges from both levels, we find

$$(q^n - 1)N_\mathcal{B} = (q^k - 1)|\mathcal{B}|. \tag{2}$$

Label each edge with the value $f(v)$, where $v \in (\mathbb{F}_q^n)^*$ is the lower node. Summing over all the edges of the graph, we obtain

$$\sum_{\mathcal{C}\in\mathcal{B}}\sum_{c\in\mathcal{C}^*} f(c) = N_\mathcal{B}\sum_{v\in(\mathbb{F}_q^n)^*} f(v) = |\mathcal{B}|\frac{q^k-1}{q^n-1}\sum_{v\in(\mathbb{F}_q^n)^*} f(v),$$

where the last equality follows from Eq. (2).

Theorem 3. *Let $\mathcal{B}$ be a balanced family of codes, and $E \subseteq \mathbb{F}_q^n$ an error set such that*

$$(q^k - 1)|E^*| < q^n - 1.$$

Then there exists a code $\mathcal{C} \in \mathcal{B}$ such that $E \cap \mathcal{C} = \{0\}$. That is, E is detectable by $\mathcal{C}$.

Proof. Let $f : \mathbb{F}_q^n \to \mathbb{C}$ be such that $f(v) = \chi_{\{v \in E^*\}}$, the indicator function for the set E^*. Applying Theorem 2, we can see that

$$\frac{q^k - 1}{q^n - 1}|E^*| = \frac{1}{q^n - 1} \sum_{v \in (\mathbb{F}_q^n)^*} f(v) = \frac{1}{|\mathcal{B}|} \sum_{\mathcal{C} \in \mathcal{B}} \sum_{c \in \mathcal{C}^*} f(c) = \frac{1}{|\mathcal{B}|} \sum_{\mathcal{C} \in \mathcal{B}} |\mathcal{C} \cap E^*|.$$

Note that for every $\mathcal{C} \in \mathcal{B}$, it's true that $|\mathcal{C} \cap E^*| \in \mathbb{N} \cup \{0\}$. By the hypothesis of our statement, we have $\frac{q^k-1}{q^n-1}|E^*| < 1$, thus

$$\frac{1}{|\mathcal{B}|} \sum_{\mathcal{C} \in \mathcal{B}} |\mathcal{C} \cap E^*| < 1.$$

Hence, by an averaging argument, it must hold that there exists some $\mathcal{C} \in \mathcal{B}$ such that $\mathcal{C} \cap E^* = \emptyset$.

Theorem 3 tells us that any error set $E \subseteq \mathbb{F}_q^n$ can be detected via an $[n, k]$ code, so long as $|E^*| < \frac{q^n-1}{q^k-1}$. If we wish to correct this error set, it suffices to consider $|\Delta E^*| < \frac{q^n-1}{q^k-1}$. Note that if $|\Delta E| < q^{n-k}$, then it can be shown that

$$|\Delta E^*| < \frac{q^n - 1}{q^k - 1}. \tag{3}$$

Theorem 3 is a generalization of the Gilbert-Varshamov bound. Indeed, if one applies Equations (1) and (3) to the set $E \subseteq B_t(0)$, we obtain the following Theorem.

Theorem 4 (Gilbert-Varshamov Bound, [29]). *Let n, k, and d be such that*

$$\sum_{i=1}^{d-1} \binom{n}{i} (q-1)^i < q^{n-k}.$$

Then there exists $\mathcal{C}$ an $[n, k]$ code $\mathcal{C}$ of minimum distance d.

It is outside the purpose of this paper to prove results on balanced families of codes, nevertheless, it is easy to show that these families exist. (Desarguesian) spreads are an example.

Definition 7. *A (Desarguesian) spread $\mathcal{S}$ is a partition of $\mathbb{F}_q^n$ into k-dimensional subspaces. That is, a spread $\mathcal{S}$ is a collection of k-dimensional subspaces such that $\bigcup_{\mathcal{C} \in \mathcal{S}} \mathcal{C} = \mathbb{F}_q^n$, and for any $\mathcal{C}_1, \mathcal{C}_2 \in \mathcal{S}$, we have $\mathcal{C}_1 \cap \mathcal{C}_2 = \{0\}$.*

It is well-known that Desarguesian spreads exist if and only if $k \mid n$. We refer the reader interested in the construction of spreads to see [24] and [25].

3.3 Density of Codes Correcting a Generic Error Set

The following theorem shows that, with high probability, a randomly chosen code corrects a fixed error set.

Theorem 5. *Let $E \subseteq \mathbb{F}_q^n$, and $k \leq n\left(1 - \frac{\log_p(|\Delta E|)}{N} - \varepsilon\right)$ for some $0 < \varepsilon < 1 - \frac{\log_p(|\Delta E|)}{N}$. Then for any $G \in \mathbb{F}_q^{k\times n}$ of rank k sampled uniformly at random, the code $\mathcal{C}$ generated from G corrects E with probability no less than $1 - q^{-n\varepsilon}$.*

Proof. Since G is sampled uniformly at random from $\mathbb{F}_q^{k\times n}$, each entry of G can be viewed as taken uniformly at random from $\mathbb{F}_q$. Then we can consider the codewords in $\mathcal{C}$ - which are linear combinations of the rows of G - as vectors with entries sampled uniformly from $\mathbb{F}_q$.

From this, the probability that an arbitrary nonzero codeword is in ΔE is

$$\frac{|(\Delta E)^*|}{|\left(\mathbb{F}_q^n\right)^*|} = \frac{|\Delta E| - 1}{q^n - 1} < \frac{|\Delta E|}{q^n} = q^{-n(1-\log_q(|\Delta E|))}.$$

Applying logarithm rules, we calculate

$$\log_q(|\Delta E|) = \log_q(|\Delta E|) = \frac{\log_p(|\Delta E|)}{\log_p(p^N)} = \frac{\log_p(|\Delta E|)}{N}.$$

Because G has rank k, there will be a total of q^k codewords in $\mathcal{C}$. Since

$$q^k q^{-n(1-\frac{\log_p(|\Delta E|)}{N})} \leq q^{n(1-\frac{\log_p(|\Delta E|)}{N}-\varepsilon)} q^{-n(1-\frac{\log_p(|\Delta E|)}{N})} = q^{-n\varepsilon},$$

we obtain that the probability a nonzero codeword from $\mathcal{C}$ will also be in ΔE is at most $q^{-n\varepsilon}$. Thus, the probability that $\Delta E \cap \mathcal{C} = \{0\}$ is bounded from below by $1 - q^{-n\varepsilon}$.

4 Generic Error SDP

With the terminology introduced at the end of Sect. 2 in mind, we can introduce the standard formulation of the Syndrome Decoding Problem.

Problem 1 (The Syndrome Decoding Problem). For an $[n, k]$ code $\mathcal{C}$ with parity-check matrix H, a syndrome vector $s \in \mathbb{F}_q^{n-k}$, and some $t \in \mathbb{N} \cup \{0\}$, find a vector $e \in \mathbb{F}_q^n$ with Hamming weight $w(e) \leq t$ such that $eH^t = s$, if such e exists.

This Hamming weight case was shown to be NP-hard for binary codes in 1978 by Berlekamp, McEliece, and Tilborg [10], then over any finite field in 1997 by Barg [8]. We are now ready to move away from the Hamming weight and define the syndrome decoding problem based on generic errors, with no specific metric.

Problem 2 (The Restricted Syndrome Decoding Problem). For an $[n, k]$ code $\mathcal{C}$ with parity-check matrix H and a syndrome vector $s \in \mathbb{F}_q^{n-k}$, find a vector $e \in \{0, \pm 1\}^n$ with Hamming weight $w(e) \leq t$ such that $eH^t = s$, if such e exists.

This sphere of vectors is the space over which [7] defines their *restricted* CVE scheme. For this paper, we no longer restrict ourselves to $\{0, \pm 1\}^n$, but instead take vectors in some arbitrary subset - though we will later apply our generic results to this specific case.

Problem 3 (The Generic Error Syndrome Decoding Problem (GE-SDP)). For an $[n, k]$ code $\mathcal{C}$ with parity-check matrix H, a syndrome vector $s \in \mathbb{F}_q^{n-k}$, and some subset $E \subseteq \mathbb{F}_q^n$, find an $e \in E$ such that $eH^t = s$, if such an e exists.

Problem 4 (Decisional GE-SDP). For an $[n, k]$ code $\mathcal{C}$ with parity-check matrix H, a syndrome vector $s \in \mathbb{F}_q^{n-k}$, and some subset $E \subseteq \mathbb{F}_q^n$, decide whether there exists an $e \in E$ such that $eH^t = s$.

This decisional GE-SDP is sometimes called the Coset Weights Problem [6]. Due to complexity theory, we know there exists a search-to-decision reduction that carries across the difficulty of the problem. Thus, the complexity of the Decisional GE-SDP will be the same as GE-SDP. We will abuse terminology and state that, for example, GE-SDP is NP-complete, despite the fact that this term applies only to the Decisional GE-SDP.

In [8], the q-ary SDP is shown to be NP-complete. More generally, over any finite ring with identity and any additive weight (i.e.: Hamming, Lee), the SDP will still be NP-complete [35]. Moreover, it is widely believed that the q-ary SDP is difficult on average, resulting in the difficulty of random instances [5].

Via a reduction argument, the authors of [7] demonstrate that the R-SDP problem is NP-complete based on the difficulty of the q-ary SDP problem. The GE-SDP presented here evidently contains all instances of SDP, as the metric is not specified. Since R-SDP is NP-complete, it represents a difficult type of problem in NP. As R-SDP is contained in our general form of SDP, we have that the hardest instances of our presentation of SDP are NP-complete, hence the GE-SDP is contained in the NP-complete complexity hierarchy. This sentiment is contained in the following theorem:

Theorem 6. *The Decisional GE-SDP is NP-complete.*

Note that this does not imply that *all* instances of the GE-SDP are NP-complete. For example, the exact computational complexity of rank-SDP, the rank metric form of the SDP, is not known [36] - but it is widely believed to be a difficult problem [19]. In practice, this rank form seems to be more difficult to solve than the Hamming weight SDP [11], and cryptographic schemes built from the rank metric appear to be more secure against decoding attacks [34]. We note that [20] demonstrates there exists a randomized polynomial time algorithm that can reduce the rank-SDP to an NP-hard problem. This is a wonderful result but not a deterministic reduction - thus, the exact complexity of the rank-SDP remains open.

5 Generic Error CVE

We now generalize the format of the CVE scheme to accept generic errors that do not depend on a specific metric. Traditionally, CVE takes an error set being the sphere of some specific Hamming weight, and samples uniformly at random a monomial transformation from $\mathfrak{M}_n$ to permute and obfuscate an error from that set [15]. In the R-SDP case from [7], this error set is taken to be $\{0, \pm 1\}^n$, with the set of monomial transformations $\widetilde{\mathfrak{M}}_n$ restricted to permit scaling factors of ± 1 only. Another variant of CVE is the one developed in [9] where the metric considered is the rank metric. Here, the error set is the set of vectors of a certain rank weight and the transformations are the natural analogue of monomial transformations in the rank-metric setting.

We note that for generic errors with no structure to speak of applying monomial transforms with no restrictions may be inappropriate. Indeed, for the set $E = \{(0, 2), (1, 0)\}$ over $\mathbb{F}_3$, the monomial transformation

$$M = \begin{pmatrix} 0 & 2 \\ 1 & 0 \end{pmatrix}.$$

would not be allowed in the generic error setting. To address this issue, we instead use the language of the stabilizer $\mathfrak{S}_E$, which may or not may not include $\mathfrak{M}_n$.

The CVE scheme based on a generic error set (GE-CVE) is shown in Fig. 2.

Remark 1. The protocol described in Fig. 2 is only one pass; in practice, many passes will be required to push the probability of error below some acceptable threshold.

We prove this system is a zero-knowledge identification scheme by showing that the following conditions hold.

- Completeness: if an honest prover and verifier follow the protocol correctly, then the verifier will always accept.
- Soundness: if an adversary outputs two valid tapes $[(c_0, c_1), z, y, b,$ and $f]$ and $[(c_0, c_1), z, y, b',$ and $f']$ with $b \neq b'$, then there exists a polynomial-time algorithm that will extract the private key e.
- Zero-Knowledge: there exists a probabilistic polynomial-time simulator that outputs tape $[(c_0, c_1), z, y, b,$ and $f]$ which is computationally indistinguishable from one produced by an honest execution.

The proofs mimic those in [7,15], and [36]. For a more rigorous treatment of these definitions, we refer the interested reader to [23] or [21].

5.1 Completeness

The hash values should match the appropriate commitment if the prover is honest.

GE-CVE

Public data: $q, n, k \in \mathbb{N}, E \subset \mathbb{F}_q^n, H \in \mathbb{F}_q^{(n-k)\times n}$

Private Key: $e \in E$

Public Key: $s = eH^t \in \mathbb{F}_q^{n-k}$

PROVER		**VERIFIER**
$u \leftarrow\!\$\, \mathbb{F}_q^n,\ M \leftarrow\!\$\, \mathfrak{S}_E$ Set $c_0 = \text{Hash}(M, uH^t)$ Set $c_1 = \text{Hash}(uM, eM)$		
	$\xrightarrow{(c_0, c_1)}$	
		$z \leftarrow\!\$\, \mathbb{F}_q^*$
	$\xleftarrow{z}$	
Set $y = (u + ze)M$		
	$\xrightarrow{y}$	
		Choose $b \in \{0, 1\}$
	$\xleftarrow{b}$	
If $b = 0$, set $f := M$ If $b = 1$, set $f := eM$		
	$\xrightarrow{f}$	
		If $b = 0$, accept if $c_0 = \text{Hash}(f, (yf^{-1})H^t - zs)$. If $b = 1$, accept if $f \in E$ and $c_1 = \text{Hash}(y - zf, f)$.

Fig. 2. One pass of the generic error CVE algorithm.

If $b = 0$, then

$$(yf^{-1})H^t - zs = ((u + ze)MM^{-1})H^t - zs = uH^t + zeH^t - zs = uH^t,$$

hence $\text{Hash}(f, (yf^{-1})H^t - zs)$ is the same as c_0, the original commitment.

On the other hand, if $b = 1$, then

$$y - zf = (u + ze)M - zeM = uM,$$

which matches commitment c_1.

5.2 Soundness

In the case of a dishonest prover, we show this adversary can convince the verifier that they are truthful with a probability that is limited by $\frac{q}{2(q-1)}$. There are two avenues to consider: the first in which the dishonest prover is expecting to receive challenge $b = 0$, and the second where they expect $b = 1$.

Call the first strategy st_0. Here, the adversary picks u and M uniformly at random, and will attempt to find e' such that $e'H^t = s$. The commitments are then

$$c_0 = \text{Hash}(M, uH^t) \text{ and } c_1 \text{ is a random string.}$$

Hence, independent of the verifier's sent value of z, the dishonest prover can respond to the challenge $b = 0$ and pass the verification test.

The second strategy st_1 is where the adversary anticipates the challenge $b = 1$. Again, u and M are chosen uniformly at random, but now they must pick $e' \in E \subseteq \mathbb{F}_q^n$. The commitments are then

$$c_0 \text{ is a random string and } c_1 = \text{Hash}(uM, e'M).$$

Since $M \in \mathfrak{S}_E$ has the property that $e'M \in E$ by definition, this is sufficient for the dishonest prover to pass the challenge for $b = 1$.

These strategies can both be improved somewhat from probability $\frac{1}{2}$ to $\frac{q}{2(q-1)}$. The dishonest prover attempts to guess the verifier's choice of z; call this guess z'. In st_0, we saw above that the adversary can correctly answer the challenge $b = 0$ independent of z, but if the guess of z' is correct they can answer the challenge $b = 1$ as well. Likewise, in st_1, they can respond to $b = 0$ if z' has been guessed correctly, and to $b = 1$ regardless of the value of z.

For one round of CVE, an adversary following strategy st_k can pick z' and thus respond correctly to the challenge b with probability

$$\mathbb{P}[b = k] + \mathbb{P}[b = 1 - k] \cdot \mathbb{P}(z' = z) = \frac{1}{2} + \frac{1}{2} \cdot \frac{1}{q-1}.$$

That is,

$$\mathbb{P}[\text{dishonest prover passes challenge } b] = \frac{q}{2(q-1)}.$$

5.3 Zero-Knowledge

In the same vein as [22], we consider a resetable, probabilistic, polynomial-time simulator. The aim of this simulation is to perform as naturally as possible, so that a third party inspecting the simulator's communication history would view it as indistinguishable from a genuine interaction.

Given that in this 5-pass scheme the verifier only ever sends information to the prover twice, they will have exactly two strategies. Indeed, let st_0 be the strategy involving taking into consideration (c_0, c_1), then producing z. Let st_1 be the other strategy of accepting (c_0, c_1) and y, then generating b as the challenge.

The simulation is executed in the following fashion:

- If $b = 0$, uniformly at random select u and M, then solve $s = e'H^t$ for e', ignoring the condition that $e' \in E$. The commitments will then be $c_0 = \text{Hash}(M, uH^t)$ with c_1 generated randomly. By simulating the verifier, the simulator will apply st_0 and return z. Computing $y = (u+ze')M$ and sending it to the simulator, it will apply st_1 and return b'.
- If $b = 1$, the simulator still chooses u and M uniformly, but now selects a random vector e', this time with $e' \in E$. The commitments will then be c_0 generated randomly and $c_1 = \text{Hash}(uM, e'M)$. Upon calling the simulator and inputting (c_0, c_1), it will apply st_0 and return z. Again, computing and sending $y = (u + ze')M$, the simulator returns b'.

The simulator then goes through the following loop: if $b' = b$, halt the simulation and output the string of communication $[(c_0, c_1), z, y, b,$ and $f]$; else, restart the protocol from the top.

After an average of $2r$ rounds, because these values are distributed uniformly, the simulator's string of communication will be indistinguishable from one that was produced from an honest execution of the protocol after r rounds - thus satisfying ZK.

We now demonstrate the $(2, 2)$-special soundness of the protocol, which tightly implies knowledge soundness.

Proposition 3. *The protocol in Fig. 2 is* $(2, 2)$*-special sound and has soundness error*

$$\frac{q}{2(q-1)}.$$

Proof. Consider the situation of an honest verifier and a cheating prover. Suppose there exist four transcripts T_1, T_2, T_3, T_4, all of which are valid and which correspond to the same commitment pair (c_0, c_1). That is, there exist $z \neq z'$ such that the prover was able to reply convincingly to queries $(z, 0)$, $(z, 1)$, $(z', 0)$ and $(z', 1)$. The commitments are then the following:

T_1: (c_0, c_1, y, M);
T_2: (c_0, c_1, y, eM);
T_3: (c_0, c_1, y', M');
T_4: (c_0, c_1, y', eM').

For the commitment c_0 to be valid for both transcripts T_1 and T_3, it must be that

$$\text{Hash}(M, (yM^{-1})H^t - zs) = c_0 = \text{Hash}(M', (y'M'^{-1})H^t - z's).$$

Therefore either there exists an extractor algorithm that can efficiently compute hash collisions, or M is indeed equal to M' and $((y - y')M^{-1})H^t = (z - z')s$.

Likewise, from the validity of transcripts T_2 and T_4, we see that

$$\text{Hash}(y - z(eM), eM) = c_1 = \text{Hash}(y' - z'(eM'), eM').$$

Recall however that $M = M'$, so either a cheating prover can find hash collisions or it holds that $y - y' = (z - z')(eM)$.

Combining these results, we find that $(e'M^{-1})H^t = s$ with $e'M^{-1} \in E$, where $e' = eM$. Thus either hash collisions have been found or this e' forms a valid key that can be used to impersonate an honest prover.

Finally, we calculate the soundness probability of Fig. 2 from [4] as

$$1 - \left(1 - \frac{1}{q-1}\right)\left(1 - \frac{1}{2}\right) = \frac{q}{2(q-1)}.$$

We end with a theorem that shows if an adversary is, in the long run, able to guess correctly more often than expected, then one of our security assumptions must have been violated.

Theorem 7. *After r rounds of the protocol in Fig. 2, if*

$$\mathbb{P}[\textit{honest verifier accepts dishonest prover}] \geq \left(\frac{q}{2(q-1)}\right)^r + \varepsilon$$

for $\varepsilon > 0$, then it is possible to either find a collision for Hash$(\cdot)$ or recover the private key e.

This is a direct result of Proposition 3. The consequence of this theorem is that either it is feasible to find collisions in the hash function, or that the GE-SDP is not an NP-complete problem - both of these violate standard cryptographic results.

6 On Polynomial Instances of GE-SDP

This section shows that if the generic error set $E \subseteq \mathbb{F}_q^n$ is included in a small Δ-closed set intersecting the code trivially, then Problem 3 can be solved by means of Gaussian elimination, leading to an attack of GE-CVE on that error set.

The attack here is a projection argument. Because the SDP was shown to be NP-complete for binary codes in [10], then over any finite field in [8], we cannot efficiently solve for an arbitrary syndrome in general.

However, solving this smaller instance over $\mathbb{F}_p$ is computationally feasible. This would not normally be of use - the errors $E \subseteq \mathbb{F}_q^n$ will generally not have any basis to speak of - but using the framework that $\overline{E}^{\Delta}$ forms a subspace allows us to find a basis. Once we recognize that we can solve this for vectors over $\mathbb{F}_p$, we can solve it for vectors over $\{0, \pm 1\}$, solving the R-SDP problem.

Recall the field $\mathbb{F}_q$, where $q = p^N$ is a prime power and that $\mathbb{F}_q$ is an N-dimensional vector space over $\mathbb{F}_p$. Thus we know that there exists an isomorphism φ mapping $\mathbb{F}_q$ into $\mathbb{F}_p^N$. If we consider the action of this isomorphism on the entries of the parity-check matrix H, we obtain a new, reduced instance - call it $H' = \varphi(H) \in \mathbb{F}_p^{N(n-k)\times n}$. The same action on the entries of s will give $s' = \varphi(s) \in \mathbb{F}_p^{N(n-k)}$. With this in mind, we need only consider the projection of the code from $\mathbb{F}_q$ down to $\mathbb{F}_p$.

Let $E \subseteq \mathbb{F}_q^n$, and consider $\overline{E}^{\Delta}$ as defined in Sect. 3. By Theorem 1, we know that $\overline{E}^{\Delta}$ is an $\mathbb{F}_p$-subspace; let

$$\overline{E}^{\Delta} = \langle E \rangle_{\mathbb{F}_p} = \langle e_1, e_2, ..., e_m \rangle_{\mathbb{F}_p}$$

where $e_1, \ldots, e_m \in E$ form a $\mathbb{F}_p$-basis.

Concerning the GE-SDP, given syndrome $s \in \mathbb{F}_q^{n-k}$ and parity check matrix $H \in \mathbb{F}_q^{(n-k)\times n}$, we can then apply Gaussian Elimination and efficiently solve the system for $\overline{E}^{\Delta}$ over $\mathbb{F}_q$:

$$s = eH^t = \lambda_1 e_1 H^t + \lambda_2 e_2 H^t + ... + \lambda_m e_m H^t. \tag{4}$$

If $\overline{E}^{\Delta} \cap \mathcal{C} = \{0\}$, since $E \subseteq \overline{E}^{\Delta}$, then E is correctable and the error $e \in E$ such that $s = eH^t$ is unique and can be found solving Eq. (4). We resume this argument in the following theorem.

Theorem 8. *Let $\mathcal{C} \subseteq \mathbb{F}_q^n$ be a code and $E \subseteq \mathbb{F}_q^n$ an error set such that $\overline{E}^{\Delta} \cap \mathcal{C} = \{0\}$. Then Problem 3 can be solved in $\mathcal{O}(n^3)$.*

For a code $\mathcal{C} \subseteq \mathbb{F}_{p^N}^n$ and an error set $E \subseteq \mathbb{F}_{p^N}^n$, for $\overline{E}^{\Delta} \cap \mathcal{C} = \{0\}$ is it necessary that

$$\dim_{\mathbb{F}_p} \overline{E}^{\Delta} + \dim_{\mathbb{F}_p} \mathcal{C} = kN + \dim_{\mathbb{F}_p} \overline{E}^{\Delta} \leq nN = \dim_{\mathbb{F}_p} \mathbb{F}_q^n.$$

This is equivalent to the condition that

$$R \leq 1 - \frac{\dim_{\mathbb{F}_p} \overline{E}^{\Delta}}{nN}. \tag{5}$$

Remark 2. Theorem 8 does not apply when $\overline{E}^{\Delta} \cap \mathcal{C} \neq \{0\}$. Indeed, in this case, the system from Eq. 4 does not have a unique solution, rather as many as $|\overline{E}^{\Delta} \cap \mathcal{C}|$. This is the case for the SDP based on the Hamming metric or on the rank metric. Indeed given a ball $E = B_r(0)$, $\overline{E}^{\Delta} = \mathbb{F}_q^n$ for either metrics.

6.1 Vulnerability of R-SDP and R-CVE

The presented attack applies to R-CVE. As already mentioned, [7] shows that the R-SDP problem is an NP-complete problem.

The attack presented in the previous section applies to all R-CVE schemes defined over a code $\mathcal{C}$ with $C \cap \mathbb{F}_q^n = \{0\}$. Indeed, the error set considered for R-CVE is $E = \{0, \pm 1\}^n$ and $\overline{E}^{\Delta} = \mathbb{F}_p^n$. Equation 5, in this case, reduces to

$$R \leq \frac{N-1}{N},$$

meaning that it is sufficient to use codes with very high rates to nullify our attack.

The cardinality of the set of codes that intersect trivially with a given error set may be calculated as a function of the q-binomial coefficient. The exact formula is outside the purpose of this paper, but it may be found as [30, Corollary 3.3].

Note that if the R-CVE is defined over a prime field, then the attack cannot be performed since $\mathcal{C} \cap \mathbb{F}_p^n = \mathcal{C}$.

Example 4. Let $p^N = 5^5$, with $n = 10$ and $k = 9$. For this example, we take $\overline{E}^{\Delta}$ to be of dimension 5 over $\mathbb{F}_5$.

From the viewpoint of rate, the inequality looks like

$$R = 0.9 \leq 1 - \frac{5}{10 \cdot 5}.$$

Hence, this code is vulnerable to being solved via basis in $\overline{E}^{\Delta}$. This example highlights that the weakness only cares about the exponent in the prime power of the code rather than the specific prime used. One can readily see that p does not appear in the inequality. Taking $p = 5$ results in the same inequality as $p = 7$, or indeed any prime.

Example 5. Consider the exact same parameters as before, except now let $\overline{E}^{\Delta}$ be slightly larger, of dimension 6 over $\mathbb{F}_5$.

Now the inequality - which does not hold - looks like

$$R = 0.9 \not\leq 0.88 = 1 - \frac{6}{10 \cdot 5}.$$

For these values, the initial conditions are not met, so the vulnerability described above does not apply here. This highlights a more general fact: when $\dim_{\mathbb{F}_p} \overline{E}^{\Delta}$ is small, the rate R will have more flexibility in the values it can take.

7 Conclusions

We have generalized the SDP and CVE to accept an error set without structure and argued the complexity of these problems. Using the set difference operation, we have constructed a particularly generic notion of detectability and correctability and applied them to this GE-SDP and GE-CVE. This also results in a generalization of the Gilbert-Varshamov Bound. It was used to give conditions that determine when there exists a code that can correct a given error set and bounds on the probability that a random code will correct an arbitrary error set. This framework demonstrates that certain GE-SDP parameters have vulnerabilities, permitting an adversary to correct the errors that are crucial to the security of the problem. We have shown that this vulnerability is only applicable when the parameters satisfy a certain bound. To conclude, we have demonstrated a vulnerability in the GE-SDP, and thus R-SDP, and presented a method of working around this susceptibility.

In regard to future work, we would welcome concrete results about the average-case complexity of GE-SDP. Seeing as R-SDP is a special case of GE-SDP with error set $\{0, \pm 1\}^n$, it is possible that other choices of small error sets may result in a practical cryptosystem. These would in turn, result in special cases of GE-CVE, which may improve the scheme.

Additionally, many of these results may be improved with the use of a trusted helper or vector (see [12,13,16]) or by leveraging the "MPC in the head" paradigm (see [2,17,18]). We relegate this to future work.

Acknowledgements. We would like to thank Violetta Weger and Paolo Santini for their helpful discussions, and Frank Kschischang for sharing some fundamental resources.

References

1. NIST's Post-Quantum Cryptography Standardization Project. https://csrc.nist.gov/Projects/post-quantum-cryptography/post-quantum-cryptography-standardization
2. Aguilar-Melchor, C., Gama, N., Howe, J., Hülsing, A., Joseph, D., Yue, D.: The return of the SDitH. Cryptology ePrint Archive, Paper 2022/1645 (2022). https://eprint.iacr.org/2022/1645
3. Albrecht, M., et al.: Classic McEliece: conservative code-based cryptography: cryptosystem specification (2022), https://classic.mceliece.org/nist.html
4. Attema, T., Fehr, S., Klooß, M.: Fiat-Shamir transformation of multi-round interactive proofs. Cryptology ePrint Archive, Paper 2021/1377 (2021). https://eprint.iacr.org/2021/1377
5. Augot, D., Finiasz, M., Sendrier, N.: A fast provably secure cryptographic hash function. IACR Cryptology ePrint Archive 2003, 230 (2003)
6. Baldi, M., Barenghi, A., Chiaraluce, F., Pelosi, G., Santini, P.: A finite regime analysis of information set decoding algorithms. Algorithms **12**(10) (2019). https://doi.org/10.3390/a12100209, https://www.mdpi.com/1999-4893/12/10/209
7. Baldi, M., et al.: A new path to code-based signatures via identification schemes with restricted errors. arXiv:abs/2008.06403 (2020)
8. Barg, A.: Some new NP-complete coding problems. Problemy Peredachi Informatsii **30**(3), 23–28 (1994)
9. Bellini, E., Caullery, F., Hasikos, A., Manzano, M., Mateu, V.: Code-based signature schemes from identification protocols in the rank metric. In: Camenisch, J.,

Papadimitratos, P. (eds.) CANS 2018. LNCS, vol. 11124, pp. 277–298. Springer, Cham (2018). https://doi.org/10.1007/978-3-030-00434-7_14
10. Berlekamp, E., McEliece, R., van Tilborg, H.: On the inherent intractability of certain coding problems (corresp.). IEEE Trans. Inf. Theory **24**, 384–386 (1978)
11. Bernstein, D.: Introduction to post-quantum cryptography. In: Bernstein, D.J., Buchmann, J., Dahmen, E. (eds.) Post-Quantum Cryptography. Springer, Heidelberg (2009). https://doi.org/10.1007/978-3-540-88702-7_1
12. Beullens, W.: Sigma protocols for MQ, PKP and SIS, and fishy signature schemes. Cryptology ePrint Archive, Paper 2019/490 (2019). https://eprint.iacr.org/2019/490
13. Bidoux, L., Gaborit, P.: Compact post-quantum signatures from proofs of knowledge leveraging structure for the PKP, SD and RSD problems. In: El Hajji, S., Mesnager, S., Souidi, E.M. (eds.) C2SI 2023. LNCS, vol. 13874, pp. 10–42. Springer, Cham (2023)
14. Bricout, R., Chailloux, A., Debris-Alazard, T., Lequesne, M.: Ternary syndrome decoding with large weight. Cryptology ePrint Archive, Paper 2019/304 (2019). https://eprint.iacr.org/2019/304
15. Cayrel, P.-L., Véron, P., El Yousfi Alaoui, S.M.: A Zero-Knowledge Identification Scheme Based on the q-ary Syndrome Decoding Problem. In: Biryukov, A., Gong, G., Stinson, D.R. (eds.) SAC 2010. LNCS, vol. 6544, pp. 171–186. Springer, Heidelberg (2011). https://doi.org/10.1007/978-3-642-19574-7_12 https://hal.inria.fr/hal-00674249
16. Feneuil, T., Joux, A., Rivain, M.: Shared permutation for syndrome decoding: New zero-knowledge protocol and code-based signature. Cryptology ePrint Archive, Paper 2021/1576 (2021). https://doi.org/10.1007/s10623-022-01116-1, https://eprint.iacr.org/2021/1576
17. Feneuil, T., Joux, A., Rivain, M.: Syndrome decoding in the head: shorter signatures from zero-knowledge proofs. Cryptology ePrint Archive, Paper 2022/188 (2022). https://doi.org/10.1007/978-3-031-15979-4_19, https://eprint.iacr.org/2022/188
18. Feneuil, T., Rivain, M.: Threshold linear secret sharing to the rescue of MPC-in-the-Head. Cryptology ePrint Archive, Paper 2022/1407 (2022). https://eprint.iacr.org/2022/1407
19. Gaborit, P., Ruatta, O., Schrek, J.: On the complexity of the rank syndrome decoding problem. CoRR abs/1301.1026 (2013)
20. Gaborit, P., Zémor, G.: On the hardness of the decoding and the minimum distance problems for rank codes. IEEE Trans. Inf. Theory **62**(12), 7245–7252 (2016). https://doi.org/10.1109/TIT.2016.2616127
21. Goldreich, O.: Foundations of Cryptography, vol. 1. Cambridge University Press, Cambridge (2001). https://doi.org/10.1017/CBO9780511546891
22. Goldreich, O.: Zero-knowledge twenty years after its invention. Cryptology ePrint Archive, Paper 2002/186 (2002). https://eprint.iacr.org/2002/186
23. Goldwasser, S., Micali, S., Rackoff, C.: The knowledge complexity of interactive proof systems. SIAM J. Comput. **18**(1), 186–208 (1989). https://doi.org/10.1137/0218012
24. Gorla, E., Manganiello, F., Rosenthal, J.: Spread codes and spread decoding in network coding. In: 2008 IEEE International Symposium on Information Theory. IEEE (2008). https://doi.org/10.1109/isit.2008.4595113
25. Gorla, E., Manganiello, F., Rosenthal, J.: An algebraic approach for decoding spread codes. Adv. Math. Commun. **6**(4), 443–466 (2012). https://doi.org/10.3934/amc.2012.6.443

26. Károlyi, G.: A compactness argument in the additive theory and the polynomial method. Discret. Math. **302**, 124–144 (2005)
27. Károlyi, G.: An inverse theorem for the restricted set addition in Abelian groups. J. Algebra **290**(2), 557–593 (2005). https://www.sciencedirect.com/science/article/pii/S0021869305002656
28. Loeliger, H.A.: Averaging bounds for lattices and linear codes. IEEE Trans. Inf. Theory **43**(6), 1767–1773 (1997). https://doi.org/10.1109/18.641543
29. Loeliger, H.A.: On the basic averaging arguments for linear codes. In: Blahut, R.E., Costello, D.J., Maurer, U., Mittelholzer, T. (eds.) Communications and Cryptography, pp. 251–261. Springer, Boston (1994). https://doi.org/10.1007/978-1-4615-2694-0_25
30. Ravagnani, A.: Whitney numbers of combinatorial geometries and higher-weight dowling lattices (2019). https://doi.org/10.48550/ARXIV.1909.10249
31. Ravagnani, A., Kschischang, F.: Adversarial network coding (2017). https://doi.org/10.48550/ARXIV.1706.05468
32. Roth, R.: Introduction to Coding Theory. Cambridge University Press (2006). https://doi.org/10.1017/CBO9780511808968
33. Silva, D., Kschischang, F.: Security for wiretap networks via rank-metric codes. In: 2008 IEEE International Symposium on Information Theory, pp. 176–180 (2008). https://doi.org/10.1109/ISIT.2008.4594971
34. Urivskiy, A., Johansson, T.: New technique for decoding codes in the rank metric and its cryptography applications. Probl. Inf. Trans. **38**, 237–246 (2002). https://doi.org/10.1023/A:1020369320078
35. Weger, V., Battaglioni, M., Santini, P., Horlemann-Trautmann, A.L., Persichetti, E.: On the hardness of the Lee syndrome decoding problem. Advances in Mathematics of Communications (2022). https://doi.org/10.3934/amc.2022029
36. Weger, V., Gassner, N., Rosenthal, J.: A survey on code-based cryptography (2022). https://doi.org/10.48550/ARXIV.2201.07119

PALOMA: Binary Separable Goppa-Based KEM

Dong-Chan Kim(✉), Chang-Yeol Jeon, Yeonghyo Kim, and Minji Kim

Future Cryptography Design Lab., Kookmin University, Seoul, Korea
{dckim,chjeon96,yh17,minji0022}@kookmin.ac.kr

Abstract. In this paper, we propose PALOMA, a new code-based key encapsulation mechanism, which is designed by combining an NP-hard SDP(Syndrome Decoding Problem)-based trapdoor with a binary separable Goppa code and FO(Fujisaki-Okamoto) transformation. Cryptographic schemes based on an SDP defined with a binary Goppa code have not been found to be vulnerable to critical attacks, and the FO transformation ensures IND-CCA2 security in the ROM(Random Oracle Model). The combination is highly regarded in cryptographic communities for its strong security guarantees. PALOMA has a public key size of approximately 300KB or more due to its SDP-based trapdoor nature. Furthermore, the key generation process, which involves generating the parity-check matrix of the scrambled Goppa code, is relatively slow compared to other post-quantum ciphers. However a primary role of post-quantum cryptography is to serve as an alternative to current cryptosystems that are vulnerable to quantum computing attacks. Therefore, in post-quantum cryptography, ensuring strong security guarantees is more important than efficiency. Consequently, we have designed PALOMA with a focus on conservative security guarantees, while ensuring that there is no significant degradation in application quality.

Keywords: code-based key encapsulation mechanism · Binary separable Goppa code · Patterson decoding · post-quantum cryptography

1 Introduction

In this paper, we propose PALOMA, a new code-based KEM(Key Encapsulation Mechanism), which has the following features:

- Trapdoor based on SDP with a binary separable Goppa code.
- IND-CCA2-secure KEM based on FO transformation.
- Parameter sets that ensure security strengths of 128, 192, and 256-bit.

1.1 Trapdoor

Syndrome Decoding Problem. SDP is a problem of finding the preimage vector with a specific Hamming weight for a given random binary parity-check

A. Esser and P. Santini (Eds.): CBCrypto 2023, LNCS 14311, pp. 144–173, 2023.
https://doi.org/10.1007/978-3-031-46495-9_8

matrix and a syndrome. In 1978, SDP was proven to be NP-hard because it is equivalent to the 3-dimensional matching problem [3,15]. The McEliece and Niederreiter cryptosystems are designed with a trapdoor based on SDP [22,24]. However, because the public key of an SDP-based trapdoor is a random-looking matrix, the public key is larger than that of other ciphers. Therefore, there have been attempts to reduce the size of a public key through cryptographic design using SDP-variant, such as rank metric-based SDP and quasi-cyclic code-based SDP. However, SDP-variants assume the problem's difficulty due to the lack of guaranteed NP-hardness for SDP and the insufficient maturity of security analysis.

A primary role of post-quantum cryptography is to serve as an alternative to current cryptosystems that are vulnerable to quantum computing attacks. Therefore, in post-quantum cryptography, ensuring strong security guarantees is more important than efficiency. We think the analysis method for SDP is sufficiently mature. Consequently, we have designed PALOMA based on SDP with a focus on conservative security guarantees, while ensuring that there is no significant degradation in application quality.

Niederreiter-type Code Scrambling (a.k.a. Syndrome Scrambling). In general, code-based cryptographic schemes use the information of a scrambled code $\widehat{\mathcal{C}}$, which is an equivalent code of the underlying code $\mathcal{C}$, as a public key pk, while the decoding information for $\mathcal{C}$ serves as a secret key sk. The McEliece scheme scrambles codewords, while the Niederreiter scheme scrambles syndromes. Syndrome scrambling has the advantage of being shorter than codeword scrambling and more intuitive for decoding, as syndromes serve as ciphertext. However, it has the drawback of requiring higher computational complexity in converting input plaintext into specific Hamming weight vectors. By the way, in the case of KEM, which does not involve encryption and thus no message input, this conversion process is unnecessary. Therefore, PALOMA adopts the syndrome scrambling approach.

Similar to the Niederreiter scheme, PALOMA uses the parity-check matrix $\widehat{\mathbf{H}}$ of a scrambled code $\widehat{\mathcal{C}}$ defined by $\mathbf{SHP}$. Here, $\mathbf{H}$ represents the parity-check matrix of $\mathcal{C}$, while $\mathbf{S}$ and $\mathbf{P}$ denote an invertible matrix and a permutation matrix respectively. The $\mathbf{P}$ used in PALOMA is randomly chosen. However, to reduce the size of a public key, the invertible matrix $\mathbf{S}$ is derived from the reduced row echelon form procedure applied to $\mathbf{HP}$, resulting in $\widehat{\mathbf{H}}$ being in a systematic form, denoted as $\widehat{\mathbf{H}} = [\mathbf{I}_{n-k} \mid \mathbf{M}]$. PALOMA uses the submatrix $\mathbf{M}$ of $\widehat{\mathbf{H}}$ as a public key, similar to Classic McEliece [4].

Binary Separable (not irreducible) Goppa Code. There are no critical attacks on cryptographic schemes based on an SDP defined with a binary separable Goppa code [13], for example, McEliece scheme, which is the first code-based cipher [22]. Many researchers have attempted to design code-based ciphers using various codes such as GRS and RM to enhance efficiency in terms of public key size and decryption speed. However, most of these schemes have been vulnerable to attacks due to their structural properties, and the remaining ones still require more rigorous security proofs [23,27]. Therefore, PALOMA adopts a binary separable Goppa code that has no attack even though it has been studied for a long time with a conservative perspective.

A binary separable Goppa code $\mathcal{C} = [n, k, \geq 2t+1]_2$ is defined by a support set L consisting of n distinct elements in $\mathbb{F}_{2^m}$ and a separable Goppa polynomial $g(X) \in \mathbb{F}_{2^m}[X]$ with degree t, for some integer $m > 1$. Typically, an irreducible polynomial is chosen as the Goppa polynomial, as every irreducible polynomial is separable. However, since the algorithms generating irreducible polynomials are probabilistic, i.e., not guaranteed to have constant-time complexity. For a generation in constant-time, PALOMA defines L and $g(X)$ with randomly chosen $n+t$ elements in $\mathbb{F}_{2^{13}}$ as follows: For a random 256-bit string r,

$$[\underbrace{\alpha_0, \alpha_1, \ldots, \alpha_{n-1}}_{n \text{ elements for } L}, \underbrace{\alpha_n, \ldots, \alpha_{n+t-1}}_{t \text{ elements for } g(X)}, \alpha_{n+t}, \ldots, \alpha_{2^m-1}] \leftarrow \textsc{Shuffle}(\mathbb{F}_{2^m}, r)$$

$$\Rightarrow \quad L \leftarrow [\alpha_0, \alpha_1, \ldots, \alpha_{n-1}], \quad g(X) \leftarrow \prod_{j=n}^{n+t-1} (X - \alpha_j).$$

After shuffling all $\mathbb{F}_{2^m}$ elements, the set of the first n elements is defined as a support set and the next t elements are the root of a Goppa polynomial with degree t. Note that $g(X)$ is separable but not irreducible in $\mathbb{F}_{2^{13}}[X]$, and we call the Goppa codes generated by the separable polynomial $g(X)$ totally decomposed Goppa codes [8]. The shuffling function, Shuffle, is a deterministic modification of the Fisher-Yates shuffling algorithm. It shuffles the set using a 256-bit string r. As a result, PALOMA efficiently generates a binary separable Goppa code in constant-time.

Patterson and Berlekamp-Massey are decoding algorithms commonly used for binary separable Goppa codes [2,19,25]. Patterson shows better speed performance compared to Berlekamp-Massey. However, it only operates when the Goppa polynomial $g(X)$ is irreducible. Therefore, PALOMA adapts the extended Patterson to handle cases where the Goppa polynomial is not irreducible [7].

1.2 KEM Structure

In general, IND-CCA2-secure schemes are constructed with OW-CPA-secure trapdoors and hash functions that are treated as random oracles. The FO transformation is a method for designing IND-CCA2-secure schemes, and it has been proven to be IND-CCA2-secure in ROM [12,14,29]. To achieve IND-CCA2-secure KEM, PALOMA is designed based on the implicit rejection $\mathsf{KEM}^{\not\perp} = \mathsf{U}^{\not\perp}[\mathsf{PKE}_1 = \mathsf{T}[\mathsf{PKE}_0, G], H]$, among FO-like transformations proposed by Hofheinz et al. [14]. This is combined with two modules: (1) T, which converts an OW-CPA-secure PKE_0 into an OW-PCA(Plaintext Checking Attack)-secure PKE_1, and (2) $\mathsf{U}^{\not\perp}$, which converts it into an IND-CCA2-secure KEM.

1.3 Parameter Sets

The security of PALOMA is evaluated by the number of bit computations of generic attacks to SDP as there are currently no known attacks on binary separable Goppa codes. ISD (Information Set Decoding) is the most powerful generic attack of an SDP. The complexity of ISD has been improved by modifications to the specific conditions for the information set [1,17,18,20,21,26,28] and birthday-type search algorithms. PALOMA determines the level of security strength by evaluating the computational complexity of the most effective attack.

PALOMA provides three parameter sets: PALOMA-128, PALOMA-192 and PALOMA-256, which correspond to security strength levels of 128-bit, 192-bit, and 256-bit, respectively. Each parameter set was carefully chosen to meet the following conditions, ensuring efficient implementation.

(1) Binary separable Goppa codes are defined in $\mathbb{F}_{2^{13}}$ which can be used for PALOMA-128, PALOMA-192, and PALOMA-256 simultaneously,
(2) $n + t \leq 2^{13}$ to define a support set and a Goppa polynomial,
(3) $n \equiv k \equiv t \equiv 0 \pmod{64}$ for 64-bit word-aligned implementation, and
(4) $k/n > 0.7$ to reduce the size of a public key.

The structure of this paper is as follows: In Sect. 1, the design rationale of PALOMA is discussed. Section 2 introduces the specification of PALOMA. Section 3 presents the performance analysis results, while Sect. 4 provides the security analysis results. In Sect. 5, the differences between PALOMA and Classic McEliece are explained, and the conclusion is drawn. Mathematical theories necessary for understanding PALOMA and the pseudo codes of PALOMA are included in the appendix.

2 Specification

The notations used throughout this paper are listed below.

$\{0,1\}^l$	set of all l-bit strings
$[i:j]$	integer array $[i, i+1, \ldots, j-1]$
$a_{[i:j]}$	substring $a_i \Vert a_{i+1} \Vert \cdots \Vert a_{j-1}$ of a bit string $a = a_0 \Vert a_1 \Vert \cdots$
$\mathbb{F}_q$	finite field with q elements
$\mathbb{F}_q^{m \times n}$	set of all $m \times n$ matrices over a field $\mathbb{F}_q$
$\mathbb{F}_q^l$	set of all $l \times 1$ matrices over a field $\mathbb{F}_q$, i.e., $\mathbb{F}_q^l := \mathbb{F}_q^{l \times 1}$ ($v \in \mathbb{F}_q^l$ is considered as a column vector)
0^l	zero vector with length l
v_I	subvector $(v_j)_{j \in I} \in \mathbb{F}_q^{\vert I \vert}$ of a vector $v = (v_0, v_1, \ldots, v_{l-1}) \in \mathbb{F}_q^l$
$\mathbf{I}_l$	$l \times l$ identity matrix
$\mathbf{M}_I$	submatrix $[m_{r,c}]_{c \in I}$ of a matrix $\mathbf{M} = [m_{r,c}]$ where r and c are row index and column index, respectively
$\mathbf{M}_{I \times J}$	submatrix $[m_{r,c}]_{r \in I,\ c \in J}$ of a matrix $\mathbf{M} = [m_{r,c}]$ where r and c are row index and column index, respectively
$\mathcal{P}_l$	set of all $l \times l$ permutation matrices
$x \xleftarrow{\$} X$	x randomly chosen in a set X

2.1 Parameter Sets

PALOMA consists of three parameter sets: PALOMA-128, PALOMA-192, and PALOMA-256 offering 128/192/256-bit security strength, respectively. Table 1 shows each parameter set.

Table 1. Parameter Sets of PALOMA

Parameter set	m	t	$n^\dagger$	$k^\ddagger$
PALOMA-128	13	64	3904	3072
PALOMA-192	13	128	5568	3904
PALOMA-256	13	128	6592	4928

$^\dagger\ n \leq 2^m - t$, $^\ddagger\ mt = n - k$

In Table 1, the parameter $m(= 13)$ represents the degree of a binary field extension. The binary extension field $\mathbb{F}_{2^{13}}(= \mathbb{F}_{2^m})$ used in PALOMA is defined by an irreducible polynomial $f(z) = z^{13} + z^7 + z^6 + z^5 + 1 \in \mathbb{F}_2[z]$, i.e., $\mathbb{F}_{2^{13}} = \mathbb{F}_2[z]/\langle f(z)\rangle$. The parameters t, n, and k denote the number of correctable errors, the length of a codeword, the dimension of a binary Goppa code, respectively.

2.2 Key Generation

The trapdoor of PALOMA is designed with SDP based on a scrambled code $\widehat{\mathcal{C}}$ of a binary separable Goppa code $\mathcal{C}$. The public key is the submatrix of the systematic parity-check matrix of $\widehat{\mathcal{C}}$, and the secret key is the necessary information for decoding and scrambling of $\mathcal{C}$. The key generation process of PALOMA is outlined below, and Algorithm 7 presents the pseudo code for key generation.

Step 1. *Generation of a random binary separable Goppa code* $\mathcal{C}$.

Generate a support set L in $\mathbb{F}_{2^{13}}$ and a Goppa polynomial $g(X) \in \mathbb{F}_{2^{13}}[X]$ for a Goppa code $\mathcal{C}$, and compute the parity-check matrix $\mathbf{H} \in \mathbb{F}_2^{13t \times n}$ of $\mathcal{C}$.

(1.1) Reorder elements of $\mathbb{F}_{2^{13}}$ with a random $r \in \{0,1\}^{256}$ using SHUFFLE (Algorithm 5).

$$\mathbb{F}_{2^{13}} = \underbrace{[0, 1, z, z+1, z^2, \ldots, z^{12} + \cdots + 1]}_{\text{lexicographic order}} \xrightarrow[\text{with } r]{\textsc{Shuffle}} [\alpha_0, \ldots, \alpha_{2^m-1}].$$

Note that we consider a field element $\alpha = \sum_{j=0}^{12} a_j z^j \in \mathbb{F}_{2^{13}}$ as an integer $\sum_{j=0}^{12} a_j 2^j \in \mathbb{Z}$ for using SHUFFLE.

(1.2) Set the support set $L = [\alpha_0, \ldots, \alpha_{n-1}]$, and set the separable Goppa polynomial $g(X) = \sum_{j=0}^{t} g_j X^j = \prod_{j=n}^{n+t-1}(X - \alpha_j) \in \mathbb{F}_{2^{13}}[X]$ of degree t.

(1.3) Compute the parity-check matrix $\mathbf{H} = \mathbf{ABC}$ where $\mathbf{A}, \mathbf{B}, \mathbf{C}$ are defined in Eq. (2).

(1.4) Parse $\mathbf{H}$ as a matrix in $\mathbb{F}_2^{13t\times n}$ because a Goppa code is a subfield subcode of the code, i.e., $\mathbf{H} = [h_{r,c}] \in \mathbb{F}_{2^{13}}^{t\times n} \Rightarrow \mathbf{H} := [h_0 \mid h_1 \mid \cdots \mid h_{n-1}] \in \mathbb{F}_2^{13t\times n}$, where $h_c := [h_{0,c}^{(0)} \mid \cdots \mid h_{0,c}^{(12)} \mid h_{1,c}^{(0)} \mid \cdots \mid h_{1,c}^{(12)} \mid \cdots \mid h_{t-1,c}^{(12)}]^T \in \mathbb{F}_2^{13t}$ and $h_{r,c}^{(j)} \in \mathbb{F}_2$ such that $h_{r,c} = \sum_{j=0}^{12} h_{r,c}^{(j)} z^j \in \mathbb{F}_{2^{13}}$ for $r \in [0:t]$ and $c \in [0:n]$.

Step 2. *Generation of a scrambled code $\widehat{\mathcal{C}}$ of $\mathcal{C}$.*

The parity-check matrix $\mathbf{H}$ of $\mathcal{C}$ is scrambled below.

(2.1) Reorder elements of $[0:n]$ with a random $r \in \{0,1\}^{256}$ using the SHUFFLE.

$$[0:n] = [0, 1, \ldots, n-1] \xrightarrow[\text{with } r]{\text{SHUFFLE}} [l_0, l_1, \ldots, l_{n-1}].$$

(2.2) Define $\mathbf{P} := \mathbf{P}_{0,l_0}\mathbf{P}_{1,l_1}\cdots\mathbf{P}_{n-1,l_{n-1}}$, and $\mathbf{P}_{j,l_j}$ is the $n \times n$ permutation matrix for swapping j-th column and l_j-th column (Algorithm 6).

(2.3) Compute $\mathbf{HP}$.

(2.4) Compute the reduced row echelon form $\widehat{\mathbf{H}}$ of $\mathbf{HP}$. If $\widehat{\mathbf{H}}_{[0:n-k]} \neq \mathbf{I}_{n-k}$, back to (2.1). Note that $\Pr[\widehat{\mathbf{H}}_{[0:n-k]} = \mathbf{I}_{n-k}] > 0.288788$.

(2.5) Define the invertible matrix $\mathbf{S}^{-1} := (\mathbf{HP})_{[0:n-k]} \in \mathbb{F}_2^{(n-k)\times(n-k)}$. Note that $\widehat{\mathbf{H}} = \mathbf{SHP}$.

Step 3. *Define a public key pk and a secret key sk.*

Since $\widehat{\mathbf{H}}$ is a matrix in systematic form, i.e., $\widehat{\mathbf{H}}_{[0:n-k]} = \mathbf{I}_{n-k}$, return $\widehat{\mathbf{H}}_{[n-k:n]}$, which is the submatrix of $\widehat{\mathbf{H}}$ consisting of the last k columns, as a public key pk and $(L, g(X), \mathbf{S}^{-1}, r)$ as a secret key sk.

Remark 1. $\mathbf{S}^{-1}$ can be derived from L, $g(X)$ and r. Both L and $g(X)$ are generated by a 256-bit random string r'. Therefore the secret key can be defined as a 512-bit string $r' \| r \in \{0,1\}^{512}$.

SHUFFLE parses a 256-bit random bit string $r = r_0 \| r_1 \| \cdots \| r_{255}$ as sixteen 16-bit non-negative integers $\widehat{r}_0, \ldots, \widehat{r}_{15}$ where $\widehat{r}_w = \sum_{j=0}^{15} r_{16w+15-j} 2^j < 2^{16}$ and uses each as a random integer required in the Fisher-Yates shuffle. Algorithm 5 shows the process of SHUFFLE in detail.

2.3 Encryption and Decryption

Encryption. PALOMA encryption is as follows (Algorithm 9).

Step 1. Retrieve the parity-check matrix $\widehat{\mathbf{H}} = [\mathbf{I}_{n-k} \mid \widehat{\mathbf{H}}_{[n-k:n]}]$ of the scrambled code $\widehat{\mathcal{C}}$ from the public key $pk = \widehat{\mathbf{H}}_{[n-k:n]} \in \mathbb{F}_2^{(n-k)\times k}$.

Step 2. Compute the $(n-k)$-bit syndrome $\widehat{s}(= \widehat{\mathbf{H}}\widehat{e})$ of an n-bit error vector input $\widehat{e}$ with $w_H(\widehat{e}) = t$, and return $\widehat{s}$ as the ciphertext of $\widehat{e}$.

Decryption. PALOMA decryption is as follows (Algorithm 9).

Step 1. Convert the syndrome $\widehat{s}$ of the input $\widehat{\mathcal{C}}$ into the syndrome $s(=\mathbf{S}^{-1}\widehat{s})$ of $\mathcal{C}$ by multiplying the secret key $\mathbf{S}^{-1}$.
Step 2. Recover the error vector e corresponding to s with the secret key $L, g(X)$, which are decoding information of $\mathcal{C}$. At that stage, we use the extended Patterson decoding introduced by Appendix A.3 (Algorithm 8).
Step 3. Return the error vector $\widehat{e}(=\mathbf{P}^{-1}e)$ of $\widehat{\mathcal{C}}$ obtained from e and the permutation matrix $\mathbf{P}^{-1}$ generated by the secret key r.

Figure 1 depicts these operations.

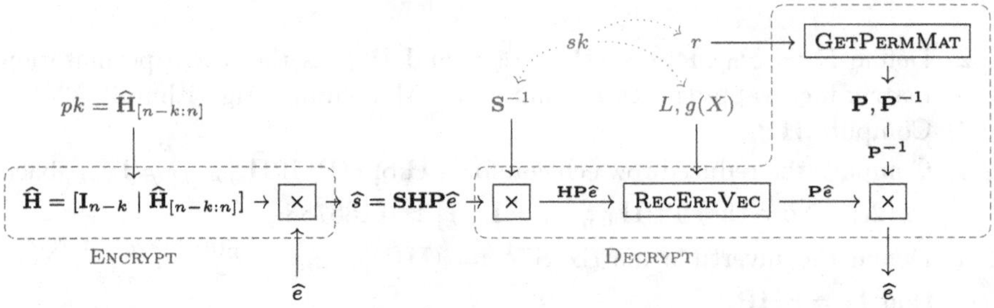

Fig. 1. PALOMA: Encryption and Decryption

2.4 Encapsulation and Decapsulation

PALOMA is a KEM designed in the random oracle model. PALOMA uses two random oracles, namely RO_G and RO_H, which are defined using the Korean KS standard hash function LSH-512 [16]. Algorithm 9 presents the definition.

Encapsulation. Encap takes a public key pk as an input and returns a key κ and the ciphertext $c = (\widehat{r}, \widehat{s})$ of κ. The procedure is as follows (Algorithm 9).

Step 1. Generate a random n-bit error vector e^* with $w_H(e^*) = t$ using GetErrVec (Algorithm 9).
Step 2. Query e^* to RO_G and obtain a 256-bit string $\widehat{r}$.
Step 3. Compute the $n \times n$ permutation $\mathbf{P}$ and its inverse $\mathbf{P}^{-1}$ corresponding to $\widehat{r}$ using GetPermMat.
Step 4. Compute $\widehat{e} = \mathbf{P}e^*$.
Step 5. Obtain the $(n-k)$-bit syndrome $\widehat{s}$ of $\widehat{e}$ using Encrypt with pk.
Step 6. Query $(e^* \| \widehat{r} \| \widehat{s})$ to RO_H and obtain a 256-bit key κ.
Step 7. Return a key κ and its ciphertext $c = (\widehat{r}, \widehat{s})$.

Decapsulation. DECAP returns the key κ when given the secret key sk and the ciphertext $c = (\widehat{r}, \widehat{s})$ as inputs. The process is as follows (Algorithm 9).

Step 1. Obtain the error vector $\widehat{e}$ by entering $\widehat{s}$ and sk into the DECRYPT.
Step 2. Generate the $n \times n$ permutation $\mathbf{P}$ and its inverse $\mathbf{P}^{-1}$ corresponding to $\widehat{r}$ which is part of the ciphertext c using GETPERMMAT.
Step 3. Compute $e^* = \mathbf{P}^{-1}\widehat{e}$.
Step 4. Query e^* to the RO_G and obtain a 256-bit string $\widehat{r}'$.
Step 5. Generate the error vector $\widetilde{e}$ using GETERRVEC with the secret key r.
Step 6. If $\widehat{r}' = \widehat{r}$, then query $(e^*\|\widehat{r}\|\widehat{s})$ to the random oracle RO_H, and if not, query $(\widetilde{e}\|\widehat{r}\|\widehat{s})$ to RO_H. Return the received bit string from RO_H as a key κ.

Figure 2 outlines ENCAP and DECAP.

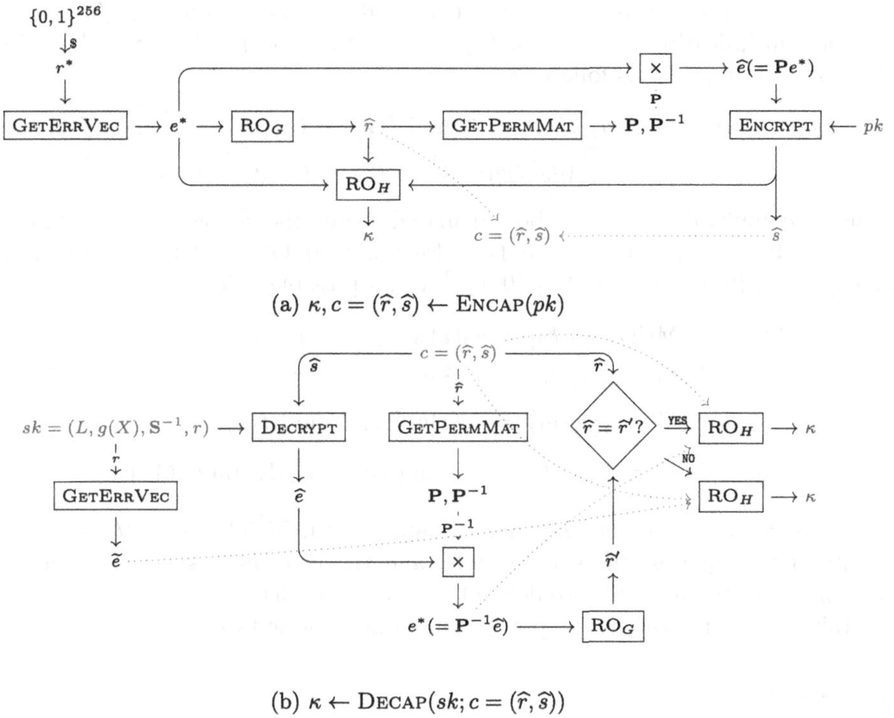

Fig. 2. PALOMA: Encapsulation and Decapsulation

3 Performance Analysis

In this section, we provide the performance analysis result of PALOMA.

3.1 Description of C Implementation

3.1.1 Data Structure for $\mathbb{F}_{2^{13}}[X]$

The elements of $\mathbb{F}_{2^{13}} = \mathbb{F}_2[z]/\langle f(z)\rangle$ are stored in the 2-byte data type `unsigned short`. The data structure for a field element $a(z) = \sum_{i=0}^{12} a_i z^i$ is defined as $0^3\|a_{12}\|\cdots\|a_0 \in \{0,1\}^{16}$. A polynomial $a(X) = \sum_{i=0}^{l} a_i X^i \in \mathbb{F}_{2^{13}}[X]$ of degree l is stored in $2(l+1)$-byte as $a_0\|\cdots\|a_l \in (\{0,1\}^{16})^{l+1}$.

3.1.2 Arithmetics in $\mathbb{F}_{2^{13}}$

PALOMA uses the pre-computed tables for multiplication, squaring, square rooting, and inversion in $\mathbb{F}_{2^{13}}$.

Multiplication. To store the multiplication of all pairs in $\mathbb{F}_{2^{13}}$, a table of 128MB (=2×2^{26}-byte) is required. In order to reduce the size of the table, PALOMA employs the multiplication of three smaller tables. Every field element $a(z) \in \mathbb{F}_{2^{13}}$ can be expressed as $a_1(z)z^7 + a_0(z)$ where $\deg(a_0) \leq 6$ and $\deg(a_1) \leq 5$. So, the multiplication of $a(z) = a_1(z)z^7 + a_0(z)$ and $b(z) = b_1(z)z^7 + b_0(z)$ in $\mathbb{F}_{2^{13}}$ can be computed as follows.

$$\begin{aligned} a(z)b(z) \bmod f(z) = &\left(a_1(z)b_1(z)z^{14} \bmod f(z)\right) + \left(a_1(z)b_0(z)z^7 \bmod f(z)\right) \\ &+ \left(a_0(z)b_1(z)z^7 \bmod f(z)\right) + \left(a_0(z)b_0(z)\right). \end{aligned}$$

Thus, the multiplication can be calculated using the following three tables $\mathsf{MUL}_{00} : \{0,1\}^7 \times \{0,1\}^7 \to \{0,1\}^{16}$, $\mathsf{MUL}_{10} : \{0,1\}^6 \times \{0,1\}^7 \to \{0,1\}^{16}$, and $\mathsf{MUL}_{11} : \{0,1\}^6 \times \{0,1\}^6 \to \{0,1\}^{16}$ for all possible pairs.

$$\begin{aligned} \mathsf{MUL}_{00}[a_0, b_0] &:= a_0(z)b_0(z) \bmod f(z), \\ \mathsf{MUL}_{10}[a_1, b_0] &:= a_1(z)b_0(z)z^7 \bmod f(z), \\ \mathsf{MUL}_{11}[a_1, b_1] &:= a_1(z)b_1(z)z^{14} \bmod f(z). \end{aligned}$$

Note that $(a_1(z)b_0(z))z^7 \bmod f(z)$ is computed using the table MUL_{10}.

Squaring, Square root, and Inversion. Tables SQU, SQRT and INV store the results of the squares, the square roots, and the inverses, respectively, for all elements in $\mathbb{F}_{2^{13}}$. Note that we define the inverse of 0 as 0.

Table 2 presents the size of pre-computed arithmetic tables.

3.2 Data Size

We determine the size of a public key pk, a secret key sk, and a ciphertext c in terms of byte strings. Each size in bytes is computed by the following formula.

$$\begin{aligned} \mathsf{bytelen}(pk) &= \mathsf{bytelen}(\widehat{\mathbf{H}}_{[n-k:n]}) = \lceil (n-k)k/8 \rceil, \\ \mathsf{bytelen}(sk) &= \mathsf{bytelen}(L) + \mathsf{bytelen}(g) + \mathsf{bytelen}(\mathbf{S}^{-1}) + \mathsf{bytelen}(r) \\ &= 2n + 2t + \lceil (n-k)^2/8 \rceil + 32, \\ \mathsf{bytelen}(c) &= \mathsf{bytelen}(\widehat{r}) + \mathsf{bytelen}(\widehat{s}) = 32 + \lceil (n-k)/8 \rceil. \end{aligned}$$

Table 2. Pre-computed Tables for Arithmetics in $\mathbb{F}_{2^{13}}$ used in PALOMA

Table	Size (in bytes)	Description
MUL_{00}	32,768 $(= 2^{14} \times 2)$	$a_0(z)b_0(z)$, $\deg(a_0), \deg(b_0) < 7$
MUL_{10}	16,384 $(= 2^{13} \times 2)$	$a_1(z)b_0(z)z^7 \bmod f(z)$, $\deg(a_1) < 6$, $\deg(b_0) < 7$
MUL_{11}	8,192 $(= 2^{12} \times 2)$	$a_1(z)b_1(z)z^{14} \bmod f(z)$, $\deg(a_1), \deg(b_1) < 6$
SQU	16,384 $(= 2^{13} \times 2)$	$a(z)^2 \bmod f(z)$, $\deg(a) < 13$
SQRT	16,384 $(= 2^{13} \times 2)$	$\sqrt{a(z)}$ where $a(z) = (\sqrt{a(z)})^2 \bmod f(z)$, $\deg(a) < 13$
INV	16,384 $(= 2^{13} \times 2)$	$a(z)^{-1}$ where $1 = a(z)^{-1}a(z) \bmod f(z)$, $\deg(a) < 13$
Total	106,496	

As stated in Remark 1, the size of a secret key can be 512-bit. However, using such a key size may adversely affect the decryption speed performance.

Table 3 shows the data size comparison among the NIST competition round 4 code-based ciphers and PALOMA.

The data size of PALOMA is similar to Classic McEliece because of the usage of SDP-based trapdoor. Compared to HQC and BIKE, the size of a public key and a secret key is relatively large. However, the size of the ciphertext which is the actual transmitted value is smaller than HQC and BIKE.

3.3 Speed

PALOMA is implemented in ANSI C. A speed benchmark was performed on the following two platforms using the GCC compiler (ver. 13.1.6) with the `-O2` optimization option:

Platform 1. macOS Monterey ver.12.5, Apple M1, 8GB RAM
Platform 2. macOS Monterey ver.12.4, Intel core i5, 8GB RAM

The results are shown in Table 4 and Table 5. Compared to Classic McEliece, an SDP-based trapdoor, PALOMA operates faster except for the parameter providing a 192-bit security. The reason is that the number of correctable errors($= t$) among 192-bit security parameters is 128 in PALOMA compared to 96 in Classic McEliece.

4 Security

4.1 OW-CPA-secure PKE = (GenKey, Encrypt, Decrypt)

When evaluating the security of PALOMA, it is important to consider that no critical attacks on binary separable Goppa codes have been reported thus far.

Table 3. Data Size Comparison of Code-based KEMs (in bytes)

Algorithm	Security	Public key	Secret key	Ciphertext	Key
hqc-128	128	2,249	40	4,481	64
BIKE	128	1,541	281	1,573	32
mceliece348864	128	261,120	6,452	128	32
PALOMA-128	128	319,488	94,496	136	32
hqc-192	192	4,522	40	9,026	64
BIKE	192	3,083	419	3,115	32
mceliece460896	192	524,160	13,568	188	32
PALOMA-192	192	812,032	355,400	240	32
hqc-256	256	7,245	40	14,469	64
BIKE	256	5,122	580	5,154	32
mceliece6688128	256	1,044,992	13,892	240	32
PALOMA-256	256	1,025,024	357,064	240	32

However, for the purpose of security analysis, we assume that the scrambled code of a Goppa code is indistinguishable from a random code. It is considered difficult to generate an effective distinguisher for Goppa codes used in PALOMA, as their rates are all less than 0.8 [10]. Therefore, the most powerful attack considered is the ISD, which is a generic attack of SDP. Consequently, the security strength of PALOMA is assessed based on the number of bit operations required for ISD process.

SDP is defined with a parity check matrix $\mathbf{H} \in \mathbb{F}_2^{(n-k)\times n}$, a syndrome $s \in \mathbb{F}_2^{n-k}$, and a Hamming weight t. We define $\mathsf{SDP}(\mathbf{H}, s, t)$ as the root set of the SDP. We also denote the set of all n-bit vectors with a Hamming weight of t as $\mathcal{E}_t^n$ and represent the zero matrix as $\mathbf{0}$. It is worth noting that the parameters n and t of PALOMA are selected to ensure that the underlying SDP possesses a unique root, and both n and t are even.

4.1.1 Assumptions for Analysis

Deterministic Fisher-Yates Shuffle Based on a 256-Bit String. PALOMA utilizes the Shuffle to generate Goppa codes, permutation matrices, and error vectors. (Algorithm 5) The deterministic Shuffle based on a 256-bit input is a modified version of the probabilistic Fisher-Yates shuffle. The Shuffle satisfies the following property.

Table 4. Speed Performance of PALOMA (in milliseconds)

		PALOMA-128		PALOMA-192		PALOMA-256	
		Plat. 1	Plat. 2	Plat. 1	Plat. 2	Plat. 1	Plat. 2
GENKEY	GENRANDGOPPACODE	15	26	74	144	93	168
	GETSCRAMBLEDCODE	42	61	179	263	211	281
	total	64	89	261	423	323	469
ENCRYPT		0.002	0.003	0.003	0.004	0.003	0.005
DECRYPT	CONSTRUCTKEYEQN	8	12	53	92	53	92
	SOLVEKEYEQN	0.2	0.4	2	3	2	3
	FINDERRVEC	1	2	3	4	4	5
	total	9	14	59	100	59	101
ENCAP		0.03	0.05	0.04	0.07	0.04	0.08
DECAP		9	15	59	101	60	101

Proposition 1. *Let* $w \in \{3904, 5568, 6592, 8192\}$. *If* $\text{SHUFFLE}([0:w], r) = \text{SHUFFLE}([0:w], \widehat{r})$ *for some* $r, \widehat{r} \in \{0,1\}^{256}$, *then* $r = \widehat{r}$.

Proof. Let $r = (r_0, r_1, \ldots, r_{15})$ and $\widehat{r} = (\widehat{r}_0, \widehat{r}_1, \ldots, \widehat{r}_{15})$ for $0 \le r_j, \widehat{r}_j < 2^{16}$. According to the nature of the Fisher-Yates shuffle, in order for the two resulting arrays to be identical, the following equation must be satisfied.

$$r_{j \bmod 16} \equiv \widehat{r}_{j \bmod 16} \pmod{w-j} \text{ for } j = 0, 1, \ldots, w-2.$$

When $w = 3904$, we obtain $r_0 \equiv \widehat{r}_0 \pmod{3904}$ and $r_0 \equiv \widehat{r}_0 \pmod{3888}$, resulting in $lcm(3904, 3888) = 948672 \mid r_0 - \widehat{r}_0$. However, since $0 \le r_0, \widehat{r}_0 < 2^{16} = 65536$, we have $r_0 = \widehat{r}_0$. By employing a similar method, we obtain $r_{15} = \widehat{r}_{15}$. The same approach applies when $w \in \{5568, 6592, 8192\}$ and yields the same result.

According to Proposition 1, SHUFFLE returns 2^{256} different arrays. PALOMA assumes that an arbitrarily selected array from the total of $w!$ possible arrays and an array obtained through SHUFFLE with an arbitrarily selected 256-bit input are indistinguishable.

Number of Equivalent Codes. PALOMA defines the support set L as the top n elements obtained by shuffling $\mathbb{F}_{2^{13}}$ using the SHUFFLE with a 256-bit input. The next t elements are defined as the roots to the Goppa polynomial $g(X)$.

Table 5. Comparison between PALOMA and Classic McEliece (in milliseconds)

		GenKey	Encap	Decap
128-bit	PALOMA-128	64	0.03	9
	mceliece348864	74	0.04	18
192-bit	PALOMA-192	258	0.04	58
	mceliece460896	211	0.06	42
256-bit	PALOMA-256	323	0.04	58
	mceliece6688128	517	0.10	82

Therefore, the expected number of equivalent codes among the 2^{256} Goppa codes generated using this method is as follows.

$$2^{256} \times \frac{\binom{2^{13}}{n} n! \binom{2^{13}-n}{t} t!}{2^{13}!} = \frac{2^{256}}{(2^{13}-n-t)!} \approx \begin{cases} 2^{-44532}, & \text{PALOMA-128}, \\ 2^{-24318}, & \text{PALOMA-192}, \\ 2^{-13117}, & \text{PALOMA-256}. \end{cases}$$

Due to the expectation values being extremely small for all three parameters of PALOMA, it is assumed that PALOMA defines the SDP using 2^{256} distinct Goppa codes.

Number of t-Hamming Weight Error Vectors. PALOMA uses GetErrVec to generate an error vector e^* with a Hamming weight of t. (Algorithm 9) In other words, based on a 256-bit sequence r^*, it shuffles the array $[0 : n]$ and defines $\text{supp}(e^*)$ as the top t elements, where $\text{supp}(e^*)$ is the non-zero position set of a given vector e^*. The expected value for the number of identical vectors among the 2^{256} error vectors generated using this method is as follows.

$$2^{256} \times \frac{\binom{n}{t} t!}{n!} = \frac{2^{256}}{(n-t)!} \approx \begin{cases} 2^{-39933}, & \text{PALOMA-128}, \\ 2^{-59410}, & \text{PALOMA-192}, \\ 2^{-72248}, & \text{PALOMA-256}. \end{cases}$$

Since the expected value for all three parameters of PALOMA is significantly smaller than 2^{-256}, it is assumed that GetErrVec returns 2^{256} distinct error vectors.

Number of Plaintexts. In PALOMA, the plaintext $\widehat{e}$ of the SDP is generated from a 256-bit string r^* through the operations GetErrVec, RO_G, and GetPermMat. (Figure 2 (a)) PALOMA assumes that the probability of having different 256-bit strings that produce the same plaintext $\widehat{e}$ through this process is extremely low and can be disregarded. In other words, PALOMA considers that there are 2^{256} possible plaintext candidates.

4.1.2 Exhaustive Search

The naive algorithm for finding roots of an SDP in PALOMA is an exhaustive search. The amount of the exhaustive search is $O\left(\binom{n}{t}(n-k)\right)$ in terms of bit operations. The amount of exhaustive search for the PALOMA parameters is as follows.

$$O\left(\binom{n}{t}(n-k)\right) \approx \begin{cases} O\left(2^{476.52}\right), & \text{PALOMA-128}, \\ O\left(2^{885.11}\right), & \text{PALOMA-192}, \\ O\left(2^{916.62}\right), & \text{PALOMA-256}. \end{cases}$$

PALOMA assumes that its underlying SDP has 2^{256} candidate roots. (Section 4.1.1) Each candidate generation requires the operations of GetErrVec, RO_G, and GetPermMat. The process of verifying if a candidate is a root requires $t-1$ $(n-k)$-bit additions. The computational cost of the Shuffle, which is the main operation in GetErrVec and GetPermMat, is negligible compared to the hash function operation, RO_G. Similarly, the computational cost of $t-1$ $(n-k)$-bit additions is also negligible compared to the RO_G operation. Therefore, the total computational cost of exhaustively searching the root candidates is $O(2^{256}T)$ where T is the computational cost of the RO_G operation. Assuming $T < 2^{40}$, generating and verifying the root candidates in PALOMA is more efficient in terms of computational cost compared to investigating all vectors with a Hamming weight of t. The set of 2^{256} root candidates can be precomputed before the start of the SDP challenge, independent of the public/secret keys. However, this requires memory of $2^{256}t\lceil\log_2 n\rceil$ bits.

4.1.3 Birthday-Type Decoding

For a random permutation matrix $\mathbf{P} \in \mathcal{P}_n$, $\mathsf{SDP}(\mathbf{H}, s, t)$ and $\mathsf{SDP}(\mathbf{HP}, s, t)$ have the necessary and sufficient conditions: $e \in \mathsf{SDP}(\mathbf{H}, s, t)$ if and only if $\mathbf{P}^{-1}e \in \mathsf{SDP}(\mathbf{HP}, s, t)$. Let $I := [0 : \frac{n}{2}]$, $J := [\frac{n}{2} : n]$, and $\widehat{\mathbf{H}} := \mathbf{HP}$. Birthday-type decoding transforms SDP until finding the root $\widehat{e} = (\widehat{e}_I \| \widehat{e}_J) := \mathbf{P}^{-1}e \in \mathsf{SDP}(\widehat{\mathbf{H}}, s, t)$ that satisfies $w_H(\widehat{e}_I) = w_H(\widehat{e}_J) = \frac{t}{2}$ for a random permutation matrix $\mathbf{P}$. To find $\widehat{e}_I$ and $\widehat{e}_J$, we check the intersection of $T_I := \{s + \widehat{\mathbf{H}}_I\widehat{e}_I \in \mathbb{F}_2^{n-k} : \widehat{e}_I \in \mathcal{E}_{t/2}^{n/2}\}$ and $T_J := \{\widehat{\mathbf{H}}_J\widehat{e}_J \in \mathbb{F}_2^{n-k} : \widehat{e}_J \in \mathcal{E}_{t/2}^{n/2}\}$. The probability that the two sets, T_I and T_J, have an intersection for a randomly chosen permutation matrix $\mathbf{P}$ is $p = \binom{n/2}{t/2}^2 / \binom{n}{t}$. Therefore, the process of transforming SDP with a new $\mathbf{P}$ must be repeated at least $1/p$ times. Algorithm 1 shows this attack in detail.

Since the number of bit computations for calculating $\widehat{\mathbf{H}}_I\widehat{e}_I$ and $\widehat{\mathbf{H}}_J\widehat{e}_J$ are $O(\binom{n/2}{t/2}(n-k))$, the total amount of computations for the PALOMA parameters is as follows.

$$O\left(2\binom{n}{t}(n-k) \Big/ \binom{n/2}{t/2}\right) = \begin{cases} O(2^{245.77}), & \text{PALOMA-128}, \\ O(2^{450.81}), & \text{PALOMA-192}, \\ O(2^{466.57}), & \text{PALOMA-256}. \end{cases}$$

Algorithm 1. Finding a root of SDP: Birthday-type Decoding

Input: $\mathbf{H} \in \mathbb{F}_2^{(n-k)\times n}$ and $s \in \mathbb{F}_2^{n-k}$, w and $I = [0 : \frac{n}{2}]$, $J = [\frac{n}{2} : n]$
Output: $e \in \mathbb{F}_2^n$ such that $\mathbf{He} = s$ and $w_H(e) = t$

while true do
  $\mathbf{P} \xleftarrow{\$} \mathcal{P}_n$
  $\widehat{\mathbf{H}} \leftarrow \mathbf{HP}$
  $T[j] \leftarrow$ null for all $j \in \{0,1\}^{n-k}$
  for $\widehat{e}_I$ in $\mathcal{E}_{t/2}^{n/2}$ **do** ▷ exhaustive search
    $u \leftarrow s + \widehat{\mathbf{H}}_I \widehat{e}_I$ // num. of bit operations $= n - k$
    $T[u] \leftarrow \widehat{e}_I$
  end for
  for $\widehat{e}_J$ in $\mathcal{E}_{t/2}^{n/2}$ **do** ▷ exhaustive search
    $u \leftarrow \widehat{\mathbf{H}}_J \widehat{e}_J$ // num. of bit operations $= n - k$
    if $T[u] \neq$ null **then**
      $\widehat{e} \leftarrow (T[u] \| \widehat{e}_J)$ ▷ $T[u] = \widehat{e}_I$
      return $\mathbf{P}\widehat{e}$
    end if
  end for
end while

To increase the probability p to a value close to 1 in birthday-type decoding, define the two subsets $I = \left[0 : \frac{n}{2} + \varepsilon\right]$ and $J = \left[\frac{n}{2} - \varepsilon : n\right]$ for some $\varepsilon > 0$. When we find $e_1, e_2 \in \mathcal{E}_{t/2}^{\frac{n}{2}+\varepsilon}$ that satisfy $s + \widehat{\mathbf{H}}_I e_1 = \widehat{\mathbf{H}}_J e_2$, it cannot be assumed that $(e_1 \| 0^{\frac{n}{2}-\varepsilon}) + (0^{\frac{n}{2}-\varepsilon} \| e_2)$ is a root. If $w_H((e_1 \| 0^{\frac{n}{2}-\varepsilon}) + (0^{\frac{n}{2}-\varepsilon} \| e_2)) = t$, then $(e_1 \| 0^{\frac{n}{2}-\varepsilon}) + (0^{\frac{n}{2}-\varepsilon} \| e_2)$ is the root. Therefore it is necessary to include this discriminant. In this attack, ε is set to a value that makes the probability $p = \binom{n/2+\varepsilon}{t/2}^2 / \binom{n}{t}$ close to 1. The calculated amount of birthday-type decoding for the PALOMA parameters is counted as follows.

$$O\left(2(n-k)\binom{n/2+\varepsilon}{t/2}\right) \approx O\left(2(n-k)\sqrt{\binom{n}{t}}\right) = \begin{cases} O(2^{244.11}), & \text{PALOMA-128}, \\ O(2^{448.91}), & \text{PALOMA-192}, \\ O(2^{464.66}), & \text{PALOMA-256}. \end{cases}$$

We consider the computation cost of this approach as a birthday-type decoding calculation, even though the overall computational complexity decreases by only about 2 or 3 bits compared to the increase in memory complexity.

4.1.4 Improved Birthday-Type Decoding

By defining two smaller SDPs from the SDP, and obtaining the roots of each SDP through birthday-type decoding, it is possible to find the root of the SDP while checking if the root candidate satisfies certain conditions. This is referred to as improved birthday-type decoding.

Consider $\mathbf{H} = \binom{\mathbf{H}_1}{\mathbf{H}_2} \in \mathbb{F}_2^{(n-k)\times n}$ as a concatenation of two submatrices, $\mathbf{H}_1 \in \mathbb{F}_2^{r\times n}$ and $\mathbf{H}_2 \in \mathbb{F}_2^{(n-k-r)\times n}$, where $r \leq n-k$. For the n-bit roots $x \in \mathsf{SDP}\left(\mathbf{H}_1, s_{[0:r]}, t/2 + \varepsilon\right)$ and $y \in \mathsf{SDP}\left(\mathbf{H}_1, 0^r, t/2 + \varepsilon\right)$ for $\mathbf{H}_1$, if x and y satisfy $\mathbf{H}_2(x+y) = s_{[r:n-k]}$ and $w_H(x+y) = t$, then $x + y \in \mathsf{SDP}(\mathbf{H}, s, t)$. Note that $|\mathsf{SDP}\left(\mathbf{H}_1, s_{[0:r]}, t/2+\varepsilon\right)| \approx |\mathsf{SDP}\left(\mathbf{H}_1, 0^r, t/2+\varepsilon\right)| \approx \frac{\binom{n}{t/2+\varepsilon}}{2^r}$. Algorithm 2 shows this method in detail.

Algorithm 2. Finding a root of SDP: Improved Birthday-type Decoding

Input: $\mathbf{H} \in \mathbb{F}_2^{(n-k)\times n}$, $s \in \mathbb{F}_2^{n-k}$, t and r
Output: $e \in \mathbb{F}_2^n$ such that $\mathbf{H}e = s$ and $w_H(e) = t$

$T[j] \leftarrow \emptyset$ for all $j \in \{0,1\}^{n-k-r}$
for x in $\mathsf{SDP}(\mathbf{H}_1, s_{[0:r]}, t/2+\varepsilon)$ **do** ▷ birthday-type decoding
  $u \leftarrow s_{[r:n-k]} + \mathbf{H}_2 x$ // num. of bit operations $= (t/2+\varepsilon)(n-k-r)$
  $T[u] \leftarrow T[u] \cup \{x\}$
end for
for y in $\mathsf{SDP}(\mathbf{H}_1, 0^r, t/2+\varepsilon)$ **do** ▷ birthday-type decoding
  $u \leftarrow \mathbf{H}_2 y$ // num. of bit operations $= (t/2+\varepsilon)(n-k-r)$
  for x in $T[u]$ **do** // $|T[u]| \approx \frac{\binom{n}{t/2+\varepsilon}}{2^r} \times \frac{1}{2^{n-k-r}}$
    $e \leftarrow x + y$ // num. of bit operations $= \frac{\binom{n}{t/2+\varepsilon}}{2^r} \times \frac{n\binom{n}{t/2+\varepsilon}}{2^{n-k}}$
    if $w_H(e) = t$ **then**
      return e
    end if
  end for
end for

The number of bit operations in this algorithm is as follows.

$$4r\sqrt{\binom{n}{t/2+\varepsilon}} + \frac{\binom{n}{t/2+\varepsilon}}{2^r}\left((t+2\varepsilon)(n-k-r) + \frac{n\binom{n}{t/2+\varepsilon}}{2^{n-k}}\right).$$

Choice of ε. When two subsets A and B with the number of elements $t/2+\varepsilon$ are randomly selected from the set $[0:n]$, the expected value $E[|A \cap B|]$ is $\frac{(t/2+\varepsilon)^2}{n}$. Therefore, for the roots x and y of each small SDP, $E[w_H(x+y)]$ is as follows.

$$\begin{aligned} E[w_H(x+y)] &= E[2(|\mathrm{supp}(x)| - |\mathrm{supp}(x) \cap \mathrm{supp}(y)|)] \\ &= 2E[|\mathrm{supp}(x)|] - 2E[|\mathrm{supp}(x) \cap \mathrm{supp}(y)|] \\ &= 2(t/2+\varepsilon) - \frac{2(t/2+\varepsilon)^2}{n}. \end{aligned}$$

Set ε to satisfy $\varepsilon = \frac{(t/2+\varepsilon)^2}{n}$, i.e., $\varepsilon = \frac{\sqrt{n^2-2nt}+(n-t)}{2}$. Then $E[w_H(x+y)] = t$.

Choice of r. For $e \in \mathsf{SDP}(\mathbf{H}, s, t)$, the number of (x, y) pairs satisfying $e = x + y$ is $|\{(x,y) \in (\mathcal{E}^n_{t/2+\varepsilon})^2 : e = x + y\}| = \binom{t}{t/2}\binom{n-t}{\varepsilon}$. Therefore, set r to satisfy $2^r \approx \binom{t}{t/2}\binom{n-t}{\varepsilon}$ to count the number of roots of small SDP accurately. The required amount of bit operations of improved birthday-type decoding for PALOMA parameters is as follows.

$$\begin{cases} O(2^{225.45})\ (\varepsilon = 3840,\ r = 61), & \mathsf{PALOMA\text{-}128}, \\ O(2^{398.84})\ (\varepsilon = 5440,\ r = 125), & \mathsf{PALOMA\text{-}192}, \\ O(2^{414.76})\ (\varepsilon = 6464,\ r = 125), & \mathsf{PALOMA\text{-}256}. \end{cases}$$

4.1.5 Information Set Decoding

ISD is a generic decoding algorithm for random linear codes. The first phase of ISD involves transforming the parity check matrix $\mathbf{H}$ into a systematic form to facilitate the identification of an error-free information set. In the second phase, error vectors satisfying specific conditions are identified, utilizing a combination of birthday attack-type searches and partial brute force attacks. Initially proposed by E. Prange in 1962, ISD has demonstrated improved computational complexity by modifying the error vector conditions and incorporating search techniques inspired by birthday attacks [1,5,6,9,11,17,18,20,21,26,28].

Procedure. ISD utilizes Proposition 2, which describes the relationship between the code $\mathcal{C}$ and the scrambled code $\widehat{\mathcal{C}}$ of $\mathcal{C}$ in terms of the root of SDP.

Proposition 2. *Let $e \in \mathsf{SDP}(\boldsymbol{H}, s, t)$. For an invertible matrix $\boldsymbol{S} \in \mathbb{F}_2^{(n-k)\times(n-k)}$ and a permutation matrix $\boldsymbol{P} \in \mathcal{P}_n$, we know $\boldsymbol{P}^{-1}e \in \mathsf{SDP}(\boldsymbol{SHP}, \boldsymbol{S}s, t)$.*

ISD is a probabilistic algorithm that modifies SDP until it finds a root that satisfies certain conditions. ISD proceeds to the following two phases.

Phase 1. Redefining a problem: Find $e \in \mathsf{SDP}(\mathbf{H}, s, t) \Rightarrow$ Find $\widehat{e} = \mathbf{P}^{-1}e \in \mathsf{SDP}(\widehat{\mathbf{H}} = \mathbf{SHP}, \widehat{s} = \mathbf{S}s, t)$ where $\mathbf{SHP} = \left(\begin{smallmatrix} \mathbf{I}_l & \mathbf{M}_1 \\ \mathbf{0} & \mathbf{M}_2 \end{smallmatrix}\right)$ is a partially systematic matrix obtained by applying elementary row operations.

Phase 2. Find $\widehat{e}(= \mathbf{P}^{-1}e) \in \mathsf{SDP}(\widehat{\mathbf{H}}, \widehat{s}, t)$ that satisfies the specific Hamming weight condition and return $e(= \mathbf{P}\widehat{e})$. If no root satisfies the condition, go back to (Phase 1).

Computational Complexity. Let p be the probability that the root $\widehat{e}$ satisfies a specific Hamming weight condition in the modified problem. The computational complexity of the ISD is $\frac{1}{p} \times$ ((Phase 1)'s computational amount + (Phase 2)'s computational amount). (Phase 1) involves modifying the problem using the Gaussian elimination. Most ISD algorithms require $O((n-k)^2 n)$ bit operations in this phase. ISD has been developed by improving the computational efficiency of (Phase 2) and the probability p.

We considered the BJMM-ISD to be the most effective ISD because the subsequent ISDs proposed after the BJMM-ISD in 2012 only provided minor improvements in specific situations [1]. Consequently, the parameters of PALOMA were chosen based on the precise calculation of the number of bit operations involved in the BJMM-ISD. The BJMM-ISD transforms the SDP into a small SDP and identifies a root of the SDP using birthday-type attacks.

Becker-Joux-May-Meurer. BJMM-ISD is an ISD that applies improved birthday-type decoding to the partial row-reduced echelon form [1]. Transform $\mathbf{H}$ into the form $\widehat{\mathbf{H}} = \left(\begin{array}{c|c} \mathbf{I}_{n-k-l} & \mathbf{H}_1 \\ \mathbf{0} & \mathbf{H}_2 \end{array}\right)$ where $\mathbf{H}_1 \in \mathbb{F}_2^{(n-k-l)\times(k+l)}$ and $\mathbf{H}_2 \in \mathbb{F}_2^{l\times(k+l)}$ by applying a partial RREF(row-reduced echelon form) operation for some $l(\leq n-k)$. For $I = [0 : n-k-l]$, $J = [n-k-l : n]$, and $L = [n-k-l : n-k]$, BJMM-ISD finds the root $\widehat{e} = (\widehat{e}_I \| \widehat{e}_J)$ of $\mathsf{SDP}(\widehat{\mathbf{H}} = \mathbf{SHP}, \widehat{s} = \mathbf{S}s, t)$ that satisfies the following conditions.

$$w_H(\widehat{e}_I) = t - p, \quad w_H(\widehat{e}_J) = p, \quad \widehat{e}_J \in \mathsf{SDP}(\mathbf{H}_2, \widehat{s}_L, p), \quad \widehat{e}_I + \widehat{e}_J \mathbf{H}_1 = \widehat{s}_I.$$

Algorithm 3. Finding a root of SDP: BJMM-ISD

Input: $\mathbf{H} \in \mathbb{F}_2^{(n-k)\times n}$, $s \in \mathbb{F}_2^{n-k}$ and t
Output: $e \in \mathbb{F}_2^n$ such that $\mathbf{H}e = s$ and $w_H(e) = t$
1: **while** true **do**
2:    $\mathbf{P} \xleftarrow{\$} \mathcal{P}_n$
3:    $\widehat{\mathbf{H}} = \mathbf{SHP} \leftarrow$ partial RREF($\mathbf{HP}$)   // num. of bit operations $= (n-k-l)(n-k)n$
4:    **if** $\widehat{\mathbf{H}}_{I\times I} = \mathbf{I}_{n-k-l}$ **then**
5:       $\mathbf{H}_1, \mathbf{H}_2 \leftarrow \mathbf{H}_{J\times I}, \mathbf{H}_{J\times L}$ $\triangleright \widehat{\mathbf{H}} = \left(\begin{array}{c|c} \mathbf{I}_{n-k-l} & \mathbf{H}_1 \\ \hline \mathbf{0} & \mathbf{H}_2 \end{array}\right)$
6:       $\widehat{s} \leftarrow \mathbf{S}s$
7:       **for** y in SDP($\mathbf{H}_2, \widehat{s}_L, p$) **do**   // $|\mathrm{SDP}(\mathbf{H}_2, \widehat{s}_L, p)| \approx \frac{\binom{k+l}{p}}{2^l}$ $\triangleright$ improved birthday-type
    decoding
8:          $x \leftarrow \widehat{s}_I + \mathbf{H}_1 y$ // num. of bit operations $= p(n-k-l)$
9:          **if** $w_H(x) = t - p$ **then**
10:             $\widehat{e} \leftarrow (x\|y)$
11:             **return** $\mathbf{P}\widehat{e}$
12:          **end if**
13:       **end for**
14:    **end if**
15: **end while**

The process of BJMM-ISD is as follows.

Phase 1. Randomly select a permutation matrix $\mathbf{P} \in \mathcal{P}_n$. Apply partial RREF to $\mathbf{HP}$ to obtain a partial canonical matrix $\widehat{\mathbf{H}} = \left(\begin{array}{c|c} \mathbf{I}_{n-k-l} & \mathbf{H}_1 \\ \hline \mathbf{0} & \mathbf{H}_2 \end{array}\right)$. In this process, the invertible matrix $\mathbf{S}$ satisfying $\widehat{\mathbf{H}} = \mathbf{SHP}$ can be obtained simultaneously. If there is no invertible matrix $\mathbf{S}$ that makes it a partial systematic form, (Phase 1) is performed again.

Phase 2. Obtain SDP($\mathbf{H}_2, \widehat{s}_L, p$) using the improved birthday-type decoding. If the root does not exist, go back to (Phase 1). If the Hamming weight of the vector $x := \widehat{s}_I + \mathbf{H}_1 y$ for $y \in$ SDP($\mathbf{H}_2, \widehat{s}_L, p$) is $t - p$, return $\mathbf{P}\widehat{e}$ because it is $\widehat{e} = (x\|y) \in$ SDP($\widehat{\mathbf{H}}, \widehat{s}, t$). If not, go back to (Phase 1).

Algorithm 3 presents the BJMM-ISD process in detail.

The probability that $\widehat{e} = \mathbf{P}^{-1}e$ satisfies the Hamming weight condition for $e \in$ SDP($\mathbf{H}, s, t$) in BJMM-ISD is as follows.

$$\Pr[\mathbf{P} \xleftarrow{\$} \mathcal{P}_n, \ (w_H(\widehat{e}_I) = t - p) \wedge (w_H(\widehat{e}_J) = p)] = \frac{\binom{n-k-l}{t-p}\binom{k+l}{p}}{\binom{n}{t}}.$$

Therefore, the calculation amount for the bit operation in the BJMM-ISD is as follows.

$$\frac{\binom{n}{t}}{\binom{n-k-l}{t-p}\binom{k+l}{p}} \left((n-k-l)(n-k)n + \frac{p(n-k-l)\binom{k+l}{p}}{2^l} + T\right),$$

where $T :=$ num. of bit operations for SDP($\mathbf{H}_2, \widehat{s}_L, p$). In this process, ε and r are set as follows for the computation of SDP($\mathbf{H}_2, \widehat{s}_L, p$).

$$\varepsilon = \frac{\sqrt{(k+l)^2 - 2(k+l)p} + (k+l-p)}{2}, \quad r = \log_2\left(\binom{p}{p/2}\binom{k+l-p}{\varepsilon}\right).$$

Table 6. Complexity of Several Attacks on PALOMA and Classic McEliece

	BJMM-ISD	Improved Birthday-type	Birthday-type	Exhaustive Search
PALOMA-128	$2^{166.21}$ $(l = 67,\ p = 14)$	$2^{225.78}$	$2^{244.11}$	$2^{476.52}$
PALOMA-192	$2^{267.77}$ $(l = 105,\ p = 22)$	$2^{399.67}$	$2^{448.91}$	$2^{885.11}$
PALOMA-256	$2^{289.66}$ $(l = 126,\ p = 26)$	$2^{415.59}$	$2^{464.66}$	$2^{916.62}$
mceliece348864	$2^{161.97}$ $(l = 66,\ p = 14)$	$2^{220.26}$	$2^{238.75}$	$2^{465.91}$
mceliece460896	$2^{215.59}$ $(l = 86,\ p = 18)$	$2^{311.80}$	$2^{345.58}$	$2^{678.88}$
mceliece6688128	$2^{291.56}$ $(l = 126,\ p = 26)$	$2^{416.95}$	$2^{466.01}$	$2^{919.32}$
mceliece6960119	$2^{289.92}$ $(l = 136,\ p = 28)$	$2^{402.41}$	$2^{443.58}$	$2^{874.57}$
mceliece8192128	$2^{318.34}$ $(l = 157,\ p = 32)$	$2^{436.05}$	$2^{484.90}$	$2^{957.10}$

The required amount of bit operations of BJMM-ISD for PALOMA parameters is as follows.

$$\begin{cases} O(2^{166.21})\ (l = 67,\ p = 14), & \text{PALOMA-128}, \\ O(2^{267.77})\ (l = 105,\ p = 22), & \text{PALOMA-192}, \\ O(2^{289.66})\ (l = 126,\ p = 26), & \text{PALOMA-256}. \end{cases}$$

Based on the above results, PALOMA claims that PALOMA-128, PALOMA-192, and PALOMA-256 have security strengths of 128-bit, 192-bit, and 256-bit, respectively. Table 6 is a comparison of the computational complexity of exhaustive search, (improved) birthday-type decoding, and BJMM-ISD for PALOMA and Classic McEliece.

4.2 IND-CCA2-Secure KEM = (GenKey, Encap, Decap)

In the IND-CCA2 security game, which stands for INDistinguishability against Adaptive Chosen-Ciphertext Attack, for the KEM = (GenKey, Encap, Decap), the challenger sends a challenge (key, ciphertext) pair to the adversary, who guesses whether the pair is correct or not. Here, "correct" means that the pair (key, ciphertext) is a valid output of the Encap. The adversary is allowed to query the Decap oracle except the challenge. We say that KEM is IND-CCA2-secure when the advantage $\mathsf{Adv}_{\mathsf{KEM}}^{\mathsf{IND\text{-}CCA2}} = |\Pr[\mathcal{A} \text{ wins IND-CCA2}] - \frac{1}{2}|$ of $\mathcal{A}$ is negligible for any probabilistic polynomial-time attacker $\mathcal{A}$. According to the analysis results in Sect. 4.1, it is assumed that the underlying PKE = (GenKey, Encrypt, Decrypt) of PALOMA is OW-CPA-secure.

PKE has the following properties. For all key pairs (pk, sk),

(1) (Injectivity) if $\textsc{Encrypt}(pk; \widehat{e}_1) = \textsc{Encrypt}(pk; \widehat{e}_2)$, then $\widehat{e}_1 = \widehat{e}_2$.
(2) (Correctness) $\Pr[\widehat{e} \neq \textsc{Decrypt}(sk; \textsc{Encrypt}(pk; \widehat{e}))] = 0$.

The Fujisaki-Okamoto transformation is a method for designing an IND-CCA2-secure scheme from an OW-CPA-secure scheme in random oracle model.

Algorithm 4. PALOMA: PKE_0, PKE_1, and $\text{KEM}^{\not\perp}$

1: **procedure** $\text{ENCRYPT}_0(pk; \hat{r}; e^*)$	1: **procedure** $\text{DECRYPT}_0(sk; c = (\hat{r}, \hat{s}))$
2: $\mathbf{P}, \mathbf{P}^{-1} \leftarrow \text{GETPERMMAT}(\hat{r})$	2: $\hat{e} \leftarrow \text{DECRYPT}(sk; \hat{s})$
3: $\hat{e} \leftarrow \mathbf{P}e^*$	3: $\mathbf{P}, \mathbf{P}^{-1} \leftarrow \text{GETPERMMAT}(\hat{r})$
4: $\hat{s} \leftarrow \text{ENCRYPT}(pk; \hat{e})$	4: $e^* \leftarrow \mathbf{P}^{-1}\hat{e}$
5: **return** $c = (\hat{r}, \hat{s})$	5: **return** e^*
6: **end procedure**	6: **end procedure**

1: **procedure** $\text{ENCRYPT}_1(pk; e^*)$	1: **procedure** $\text{DECRYPT}_1(sk; c = (\hat{r}, \hat{s}))$
2: $\hat{r} \leftarrow \text{RO}_G(e^*)$	2: $e^* \leftarrow \text{DECRYPT}_0(sk; \hat{s})$
3: $c = (\hat{r}, \hat{s}) \leftarrow \text{ENCRYPT}_0(pk; \hat{r}; e^*)$	3: $\hat{r}' \leftarrow \text{RO}_G(e^*)$
4: **return** $c = (\hat{r}, \hat{s})$	4: **if** $\hat{r} \neq \hat{r}'$ **then**
5: **end procedure**	5: **return** $\perp$
	6: **end if**
	7: **return** e^*
	8: **end procedure**

1: **procedure** $\text{ENCAP}(pk)$	1: **procedure** $\text{DECAP}(sk; c = (\hat{r}, \hat{s}))$
2: $r^* \xleftarrow{\$} \{0,1\}^{256}$	2: $e^* \leftarrow \text{DECRYPT}_1(sk; c = (\hat{r}, \hat{s}))$
3: $e^* \leftarrow \text{GETERRVEC}(r^*)$	3: **if** $e^* = \perp$ **then**
4: $c = (\hat{r}, \hat{s}) \leftarrow \text{ENCRYPT}_1(pk; e^*)$	4: $\tilde{e} \leftarrow \text{GETERRVEC}(r)$ // $r \leftarrow sk$
5: $\kappa \leftarrow \text{RO}_H(e^* \| \hat{r} \| \hat{s})$	5: $\kappa \leftarrow \text{RO}_H(\tilde{e} \| \hat{r} \| \hat{s})$
6: **return** κ and $c = (\hat{r}, \hat{s})$	6: **else**
7: **end procedure**	7: $\kappa \leftarrow \text{RO}_H(e^* \| \hat{r} \| \hat{s})$
	8: **end if**
	9: **return** κ
	10: **end procedure**

There are several variants of the Fujisaki-Okamoto transformation. Using the above properties, PALOMA is designed based on the implicit rejection $\text{KEM}^{\not\perp} = \mathsf{U}^{\not\perp}[\text{PKE}_1 = \mathsf{T}[\text{PKE}_0, G], H]$ among FO-like transformations proposed by Hofheinz et al. [14]. This is combined with two modules: T: converting OW-CPA-secure PKE_0 to OW-PCA(Plaintext-Checking Attack)-secure PKE_1 and $\mathsf{U}^{\not\perp}$: converting it to IND-CCA2-secure KEM as follows.

$$\text{OW-CPA-secure } \text{PKE}_0 = (\text{GENKEY}, \text{ENCRYPT}_0, \text{DECRYPT}_0)$$

$$\xrightarrow[\text{with a random oracle } G]{\mathsf{T}} \text{OW-PCA-secure } \text{PKE}_1 = (\text{GENKEY}, \text{ENCRYPT}_1, \text{DECRYPT}_1)$$

$$\xrightarrow[\text{with a random oracle } H]{\mathsf{U}^{\not\perp}} \text{IND-CCA2-secure} \text{KEM}^{\not\perp} = (\text{GENKEY}, \text{ENCAP}, \text{DECAP}).$$

4.2.1 OW-CPA-secure PKE_0

PKE_0 is defined with the PKE and GETPERMMAT of PALOMA as follows.

$$\text{ENCRYPT}_0(pk; \hat{r}; e^*) := (\hat{r}, \text{ENCRYPT}(pk; \mathbf{P}e^*)) \text{ where } (\mathbf{P}, \mathbf{P}^{-1}) = \text{GETPERMMAT}(\hat{r}).$$

Algorithm 4 shows the detailed process of PKE_0. As PKE is assumed to be OW-CPA-secure, it follows that PKE_0 is also OW-CPA-secure.

4.2.2 OW-PCA-secure PKE_1

The transform T for converting OW-PCA-secure PKE_0 to OW-PCA-secure PKE_1 is defined by

$$\text{ENCRYPT}_1(pk; e^*) := \text{ENCRYPT}_0(pk; G(e^*); e^*), \text{ where } G \text{ is a random oracle.}$$

Algorithm 4 shows PKE_1 constructed by this transformation T and a random oracle RO_G.

For any OW-PCA-attackers $\mathcal{B}$ on PKE_1, there exists an OW-CPA-attacker $\mathcal{A}$ on PKE_0 satisfying the inequality below [14, Theorem 3.1].

$$\mathsf{Adv}^{\mathsf{OW\text{-}PCA}}_{\mathsf{PKE}_1}(\mathcal{B}) \leq (q_G + q_P + 1)\mathsf{Adv}^{\mathsf{OW\text{-}CPA}}_{\mathsf{PKE}_0}(\mathcal{A}), \tag{1}$$

where q_G and q_P are the number of queries to the random oracle RO_G and the plaintext-checking oracle PCO, which can be implemented by re-encryption. Note that PALOMA cannot implement a ciphertext validity oracle CVO because it generates error vectors as messages from a 256-bit string. From Eq. (1), if PKE_0 is OW-CPA-secure, $\mathsf{Adv}^{\mathsf{OW\text{-}PCA}}_{\mathsf{PKE}_1}(\mathcal{B})$ is negligible, so we have PKE_1 is OW-PCA-secure.

4.2.3 IND-CCA2-secure $\mathsf{KEM}^{\not\perp}$

The transform $\mathsf{U}^{\not\perp}$ for converting OW-PCA-secure PKE_1 to IND-CCA2-secure PKE_1 is as follows.

$$\textsc{Encap}(pk) := (c = \textsc{Encrypt}_1(pk; e^*), \mathrm{RO}_H(e^* \| c)).$$

Algorithm 4 shows $\mathsf{KEM}^{\not\perp}$ of PALOMA constructed by using the transformation $\mathsf{U}^{\not\perp}$ and a random oracle RO_H.

For any IND-CCA2-attackers $\mathcal{B}$ on $\mathsf{KEM}^{\not\perp}$, there exists an OW-PCA-attacker $\mathcal{A}$ on PKE_1 satisfying the inequality below [14, Theorem 3.4].

$$\mathsf{Adv}^{\mathsf{IND\text{-}CCA2}}_{\mathsf{KEM}^{\not\perp}}(\mathcal{B}) \leq \frac{q_H}{2^{256}}\mathsf{Adv}^{\mathsf{OW\text{-}PCA}}_{\mathsf{PKE}_1}(\mathcal{A}),$$

where q_H is the number of queries to the plaintext-checking oracle. Therefore, if PKE_1 is OW-PCA-secure, $\mathsf{Adv}^{\mathsf{OW\text{-}PCA}}_{\mathsf{PKE}_1}(A)$ is negligible. Consequently, we have that $\mathsf{KEM}^{\not\perp}$ is IND-CCA2-secure.

5 Conclusion

In this paper, we introduce PALOMA, an IND-CCA2-secure KEM based on an SDP with a binary separable Goppa code. While the components and mechanisms used in PALOMA have been studied for a long time, no critical attacks have been found. Many cryptographic communities believe that the scheme constructed by these is secure. The Classic McEliece, which is the 4th round cipher of the NIST PQC competition, was also designed based on similar principles [4]. Both PALOMA and Classic McEliece have similar public key sizes. However, PALOMA is designed with a focus on deterministic algorithms for constant-time operations, making it more efficient in terms of implementation speed compared to Classic McEliece. We give the feature comparison between PALOMA and Classic McEliece in Table 7.

Table 7. Comparison between PALOMA and Classic McEliece

	PALOMA	Classic McEliece
Structure	Fujisaki-Okamoto-structure KEM (implicit rejection)	SXY-structure KEM (implicit rejection)
Problem	SDP	SDP
Trapdoor type	Niederreiter	Niederreiter
Field $\mathbb{F}_{q^m}$	$\mathbb{F}_{2^{13}}$	$\mathbb{F}_{2^{12}}, \mathbb{F}_{2^{13}}$
Linear code $\mathcal{C}$	Binary separable Goppa code	Binary irreducible Goppa code
Goppa polynomial $g(X)$	Separable (not irreducible)	Irreducible
Time for generating $g(X)$	Constant	Non-constant
Parity-check matrix $\mathbf{H}$ of $\mathcal{C}$	**ABC**	**BC**
Parity-check matrix $\widehat{\mathbf{H}}$ of $\widehat{\mathcal{C}}$	Systematic	Systematic
Decoding algorithm	Extended Patterson	Berlekamp-Massey
Probability of decryption failure (correctness)	0	0

A primary role of post-quantum cryptography is to serve as an alternative to current cryptosystems that are vulnerable to quantum computing attacks. Therefore, we have designed PALOMA with a conservative approach, and thus, we firmly believe that PALOMA can serve as a dependable alternative to existing cryptosystems in the era of quantum computers.

Acknowledgements. We are grateful to the anonymous reviewers for their help in improving the quality of the paper. This work was supported by the Ministry of Education of the Republic of Korea and the National Research Foundation of Korea (No.NRF-2021R1F1A1062305).

A Mathematical Background

In this section, we provide the necessary mathematical background to understand the operating principles of PALOMA.

A.1 Syndrome Decoding Problem

SDP is the problem of finding a syndrome preimage vector with a specific Hamming weight. The formal definition of SDP is as follows.

Definition 1. (*SDP*). *Given a parity-check matrix* $\boldsymbol{H} \in \mathbb{F}_2^{(n-k)\times n}$ *of a random binary linear code* $\mathcal{C} = [n, k]_2$, *a syndrome* $s \in \mathbb{F}_2^{n-k}$ *and* $t \in \{1, \ldots, n\}$, *find the vector* $e \in \mathbb{F}_2^n$ *that satisfies* $\boldsymbol{H}e = s$ *and* $w_H(e) = t$.

SDP has been proven to be an NP-hard problem due to its equivalence to the 3-dimensional matching problem, as demonstrated in 1978 [3,15].

Number of Roots of SDP. The preimage vector with Hamming weight less than or equal to $\lfloor \frac{d-1}{2} \rfloor$ is unique. Generally, in SDP-based schemes, the Hamming weight condition w of SDP is set to $\lfloor \frac{d-1}{2} \rfloor$ for the uniqueness of root.

A.2 Binary Separable Goppa Code

Binary separable Goppa codes are special cases of algebraic-geometric codes proposed by V. D. Goppa in 1970 [13]. The formal definition of a binary separable Goppa code over $\mathbb{F}_2$ is as follows.

Definition 2 (Binary Separable Goppa code). *For a set of distinct $n(\leq 2^m)$ elements $L = [\alpha_0, \alpha_1, \ldots, \alpha_{n-1}]$ of $\mathbb{F}_{2^m}$ and a separable polynomial $g(X) = \sum_{j=0}^{t} g_j X^j \in \mathbb{F}_{2^m}[X]$ of degree t such that none of the elements of L are roots of $g(X)$, i.e., $g(\alpha) \neq 0$ for all $\alpha \in L$, a binary separable Goppa code of length n over $\mathbb{F}_2$ is the subspace $\mathcal{C}_{L,g}$ of $\mathbb{F}_2^n$ defined by*

$$\mathcal{C}_{L,g} := \{(c_0, \ldots, c_{n-1}) \in \mathbb{F}_2^n : \sum_{j=0}^{n-1} c_j (X - \alpha_j)^{-1} \equiv 0 \pmod{g(X)}\},$$

where $(X - \alpha)^{-1}$ is the polynomial of degree $t-1$ satisfying $(X - \alpha)^{-1}(X - \alpha) \equiv 1 \pmod{g(X)}$. L and $g(X)$ are referred to as a support set and a Goppa polynomial, respectively.

Dimension and Minimum Hamming Distance. The dimension k and the minimum Hamming distance d of $\mathcal{C}_{L,g}$ satisfy $k \geq n - mt$ and $d \geq 2t + 1$. PALOMA set the dimension k of $\mathcal{C}_{L,g}$ to $n - mt$ and the Hamming weight condition of the SDP to t to ensure the uniqueness of the root.

Parity-Check Matrix. The parity-check matrix $\mathbf{H}$ of $\mathcal{C}_{L,g}$ is defined with each coefficient of the polynomial $(X - \alpha_j)^{-1}$ with degree $t-1$, and $\mathbf{H}$ can be decomposed into $\mathbf{ABC}$, defined by

$$\mathbf{A} := \begin{pmatrix} g_1 & g_2 & \cdots & g_t \\ g_2 & g_3 & \cdots & 0 \\ \vdots & \vdots & \ddots & \vdots \\ g_t & 0 & \cdots & 0 \end{pmatrix} \in \mathbb{F}_{2^m}^{t \times t}, \quad \mathbf{B} := \begin{pmatrix} \alpha_0^0 & \alpha_1^0 & \cdots & \alpha_{n-1}^0 \\ \vdots & \vdots & \ddots & \vdots \\ \alpha_0^{t-2} & \alpha_1^{t-2} & \cdots & \alpha_{n-1}^{t-2} \\ \alpha_0^{t-1} & \alpha_1^{t-1} & \cdots & \alpha_{n-1}^{t-1} \end{pmatrix} \in \mathbb{F}_{2^m}^{t \times n},$$
$$\text{and } \mathbf{C} := \begin{pmatrix} g(\alpha_0)^{-1} & 0 & \cdots & 0 \\ 0 & g(\alpha_1)^{-1} & \cdots & 0 \\ \vdots & \vdots & \ddots & \vdots \\ 0 & 0 & \cdots & g(\alpha_{n-1})^{-1} \end{pmatrix} \in \mathbb{F}_{2^m}^{n \times n}. \tag{2}$$

Since the matrix $\mathbf{A}$ is invertible ($g_t \neq 0$), $\mathbf{BC}$ is another parity-check matrix of $\mathcal{C}_{L,g}$. Classic McEliece uses $\mathbf{BC}$ as a parity-check matrix.

A.3 Extended Patterson for Binary Separable Goppa code

Patterson decoding is the algorithm for a binary irreducible Goppa code, not a separable Goppa code. However, it can be extended for a binary separable Goppa code [7,25]. Given a syndrome vector $s \in \mathbb{F}_2^{n-k}$, the extended Patterson decoding procedure to find the preimage vector $e \in \mathbb{F}_2^n$ of s with $w_H(e) = t$ is as follows. (Note that preimage vector is called an error vector in coding theory)

Step 1. Convert the syndrome vector s into the syndrome polynomial $s(X) \in \mathbb{F}_{2^m}[X]$ of degree t or less.
Step 2. Derive the key equation for finding the error locator polynomial $\sigma(X) \in \mathbb{F}_{2^m}[X]$ of degree t.
Step 3. Solve the key equation using the extended Euclidean algorithm.
Step 4. Calculate $\sigma(X)$ using a root of the key equation.
Step 5. Find all roots of $\sigma(X)$ and compute the preimage vector e. At this stage, to ensure resistance against timing attacks, PALOMA uses the exhaustive search.

In the above procedure, the error locator polynomial $\sigma(X) = \prod_{j \in \mathrm{supp}(e)}(X - \alpha_j) \in \mathbb{F}_{2^m}[X]$ and $\sigma(X)$ satisfies the following identity.

$$\sigma(X)s(X) \equiv \sigma'(X) \pmod{g(X)}. \tag{3}$$

Note that $\sigma(X)$ satisfying Eq. (3) is unique since the number of errors is t. In $\mathbb{F}_{2^m}[X]$, all polynomials $f(X)$ has two polynomials $a(X)$ and $b(X)$ such that $f(X) = a(X)^2 + b(X)^2 X$, $\deg(a) \leq \lfloor t/2 \rfloor$, and $\deg(b) \leq \lfloor (t-1)/2 \rfloor$. Thus, if $\sigma(X) = a(X)^2 + b(X)^2 X$, Eq. (3) can be rewritten as follows.

$$b(X)^2(1 + Xs(X)) \equiv a(X)^2 s(X) \pmod{g(X)}. \tag{4}$$

When $g(X)$ is irreducible, $s^{-1}(X)$ and $\sqrt{s^{-1}(X) + X}$ exist in modulo $g(X)$. Patterson decoding uses the extended Euclidean algorithm to find $a(X)$ and $b(X)$ of the following key equation to generate the error locator polynomial $\sigma(X)$.

$$b(X)\sqrt{(s^{-1}(X) + X)} \equiv a(X) \pmod{g(X)},\ \deg(a) \leq \lfloor t/2 \rfloor,\ \deg(b) \leq \lfloor (t-1)/2 \rfloor.$$

However, if $g(X)$ is separable, the existence of $s^{-1}(X)$ cannot be guaranteed because $g(X)$ and $s(X)$ are unlikely to be relatively prime.

We define

$$\widetilde{s}(X) := 1 + Xs(X), \quad g_1(X) := \gcd(g(X), s(X)), \quad g_2(X) := \gcd(g(X), \widetilde{s}(X)).$$

Since $\gcd(s(X), \widetilde{s}(X)) = \gcd(s(X), \widetilde{s}(X) \bmod s(X)) = \gcd(s(X), 1) \in \mathbb{F}_{2^m} \setminus \{0\}$, we know

$$g \mid b^2\widetilde{s} + a^2 s \overset{g_1 \mid g}{\Longrightarrow} g_1 \mid b^2\widetilde{s} + a^2 s \overset{g_1 \mid s}{\Longrightarrow} g_1 \mid b^2\widetilde{s} \overset{g_1 \nmid \widetilde{s}}{\Longrightarrow} g_1 \mid b^2 \Rightarrow g_1 \mid b,$$
$$g \mid b^2\widetilde{s} + a^2 s \overset{g_2 \mid g}{\Longrightarrow} g_2 \mid b^2\widetilde{s} + a^2 s \overset{g_2 \mid \widetilde{s}}{\Longrightarrow} g_2 \mid a^2 s \overset{g_2 \nmid s}{\Longrightarrow} g_2 \mid a^2 \Rightarrow g_2 \mid a.$$

Therefore, the following polynomials can be defined in $\mathbb{F}_{2^m}[X]$.

$$b_1(X) := \frac{b(X)}{g_1(X)}, \quad a_2(X) := \frac{a(X)}{g_2(X)}, \quad g_{12}(X) := \frac{g(X)}{g_1(X)g_2(X)},$$
$$\widetilde{s}_2(X) := \frac{\widetilde{s}(X)}{g_2(X)}, \quad s_1(X) := \frac{s(X)}{g_1(X)}.$$

Equation (4) can be rewritten as follows.

$$b(X)^2\widetilde{s}(X) \equiv a(X)^2 s(X) \pmod{g(X)}$$
$$\Rightarrow\ b_1^2(X)g_1(X)\widetilde{s}_2(X) \equiv a_2^2(X)g_2(X)s_1(X) \pmod{g_{12}(X)}.$$

Because $\gcd(g_2(X), g_{12}(X)), \gcd(s_1(X), g_{12}(X))$ is an element of $\mathbb{F}_{2^m}$, we know $\gcd(g_2(X)s_1(X), g_{12}(X)) \in \mathbb{F}_{2^m}$. Therefore, there exists the inverse of $g_2(X)s_1(X)$ modulo $g_{12}(X)$, and we have the following equation.

$$b_1^2(X)u(X) \equiv a_2^2(X) \pmod{g_{12}(X)} \text{ where } u(X) := g_1(X)\widetilde{s}_2(X)(g_2(X)s_1(X))^{-1}.$$

Since $u(X)$ has a square root modulo $g_{12}(X)$ (Remark 2), $a(X) = a_2(X)g_2(X)$ and $b(X) = b_1(X)g_1(X)$ are obtained by calculating $a_2(X)$ and $b_1(X)$ that satisfy the following key equation using the extended Euclidean algorithm.

$$b_1(X)\sqrt{u(X)} \equiv a_2(X) \pmod{g_{12}(X)},$$
$$\deg(a_2) \le \lfloor t/2 \rfloor - \deg(g_2),\ \deg(b_1) \le \lfloor (t-1)/2 \rfloor - \deg(g_1).$$

Remark 2. Since all elements of $\mathbb{F}_{2^{13}}$ are roots of the equation $X^{2^{13}} - X = 0$ and $g_{12}(X) \mid X^{2^{13}} - X$, we know $\sqrt{X} = X^{2^{12}} \bmod g_{12}(X)$. A polynomial $u(X) = \sum_{i=0}^{l} u_i X^i \in \mathbb{F}_{2^{13}}[X]$ of degree l can be written as $u(X) = (\sum_{i=0}^{\lfloor l/2 \rfloor} \sqrt{u_{2i}} X^i)^2 + (\sum_{i=0}^{\lfloor (l-1)/2 \rfloor} \sqrt{u_{2i+1}} X^i)^2 X$ where $\sqrt{u_j} = (u_j)^{2^{12}}$ for all j. Thus, the square root $\sqrt{u(X)}$ of $u(X)$ modulo $g_{12}(X)$ is

$$\sqrt{u(X)} = \left(\sum_{i=0}^{\lfloor l/2 \rfloor} \sqrt{u_{2i}} X^i\right) + \left(\sum_{i=0}^{\lfloor (l-1)/2 \rfloor} \sqrt{u_{2i+1}} X^i\right) \sqrt{X} \bmod g_{12}(X).$$

B Pseudo codes for PALOMA

In this section, we provide pseudo codes of the functions used in PALOMA.

Algorithm 5. Shuffling with a 256-bit Seed

Input: An ordered set $A = [A_0, A_1, \ldots, A_{l-1}]$ and a 256-bit seed $r = r_0 \| \cdots \| r_{255}$
Output: A shuffled set A

```
procedure SHUFFLE(A, r)
    r̂_w ← Σ_{j=0}^{15} r_{16w+15−j} 2^j for w = 0, 1, ..., 15        ▷ r̂_w ∈ {0, 1, ..., 2^16 − 1}
    w ← 0
    for i ← l − 1 downto 1 do
        j ← r̂_{w mod 16} mod i + 1
        swap(A_i, A_j)
        w ← w + 1
    end for
    return A
end procedure
```

Algorithm 6. Generation of a n-bit Permutation with a 256-bit Seed

Input: A 256-bit seed r
Output: An $n \times n$ permutation matrix $\mathbf{P}, \mathbf{P}^{-1}$

procedure GetPermMat(r)
  $[l_0, l_1, \ldots, l_{n-1}] \leftarrow$ Shuffle($[0:n], r$) ▷ Alg. 5
  $\mathbf{P} \leftarrow \prod_{j=0}^{n-1} \mathbf{P}_{j,l_j} = \mathbf{P}_{0,l_0}\mathbf{P}_{1,l_1} \cdots \mathbf{P}_{n-1,l_{n-1}}$ where $\mathbf{P}_{i,j}$ is the $n \times n$ permutation matrix for swapping i-th column and j-th column.
  $\mathbf{P}^{-1} \leftarrow \mathbf{P}_{n-1,l_{n-1}} \cdots \mathbf{P}_{1,l_1}\mathbf{P}_{0,l_0}$
  return $\mathbf{P}, \mathbf{P}^{-1}$
end procedure

Algorithm 7. Generation of Key Pair

Input: Parameter set (t, n)
Output: A public key pk and a secret key sk

procedure GenKey(t, n)
  $L, g(X), \mathbf{H} \leftarrow$ GenRandGoppaCode(t, n)
  $\mathbf{S}^{-1}, r, \widehat{\mathbf{H}} \leftarrow$ GetScrambledCode($\mathbf{H}$) ▷ $\widehat{\mathbf{H}} = \mathbf{SHP}$
  $pk, sk \leftarrow \widehat{\mathbf{H}}_{[n-k:n]}, (L, g(X), \mathbf{S}^{-1}, r)$
  return pk and sk
end procedure

Input: Parameter set (t, n)
Output: A support set L, a Goppa poly. $g(X)$ and a parity-check matrix $\mathbf{H}$ of $\mathcal{C}_{L,g}$

procedure GenRandGoppaCode(t, n)
  $r \xleftarrow{\$} \{0,1\}^{256}$
  $[\alpha_0, \ldots, \alpha_{2^{13}-1}] \leftarrow$ Shuffle($\mathbb{F}_{2^{13}}, r$) ▷ Alg. 5
  $L, g(X) \leftarrow [\alpha_0, \ldots, \alpha_{n-1}], \prod_{j=n}^{n+t-1}(X - \alpha_j)$
  $\mathbf{H} = [h_{r,c}] \leftarrow \mathbf{ABC} \in \mathbb{F}_{2^{13}}^{t \times n}$ ▷ $\mathbf{A}, \mathbf{B}, \mathbf{C}$ are defined in Eq. (2)
  $h_c \leftarrow [h_{0,c}^{(0)} \mid \cdots \mid h_{0,c}^{(12)} \mid h_{1,c}^{(0)} \mid \cdots \mid h_{1,c}^{(12)} \mid \cdots \mid h_{t-1,c}^{(12)}]^T \in \mathbb{F}_2^{13t}$ for $c \in [0:n]$ where $h_{r,c}^{(j)} \in \mathbb{F}_2$ such that $h_{r,c} = \sum_{j=0}^{12} h_{r,c}^{(j)} z^j \in \mathbb{F}_{2^{13}}$
  $\mathbf{H} \leftarrow [h_0 \mid h_1 \mid \cdots \mid h_{n-1}] \in \mathbb{F}_2^{13t \times n}$
  return $L, g(X), \mathbf{H}$
end procedure

Input: A parity-check matrix $\mathbf{H}$ of $\mathcal{C}$
Output: An invertible matrix $\mathbf{S}^{-1}$, a random bit string r and a parity-check matrix $\widehat{\mathbf{H}}$ of $\widehat{\mathcal{C}}$

1: **procedure** GetScrambledCode($\mathbf{H}$)
2:   $r \xleftarrow{\$} \{0,1\}^{256}$
3:   $\mathbf{P}, \mathbf{P}^{-1} \leftarrow$ GetPermMat(r) ▷ Alg. 6
4:   $\widehat{\mathbf{H}} \leftarrow$ Rref($\mathbf{HP}$)
5:   **if** $\widehat{\mathbf{H}}_{[0:n-k]} \neq \mathbf{I}_{n-k}$ **then**
6:     Go back to line 2.
7:   **end if**
8:   $\mathbf{S}^{-1} \leftarrow (\mathbf{HP})_{[0:n-k]}$
9:   **return** $\mathbf{S}^{-1}, r, \widehat{\mathbf{H}}$
10: **end procedure**

Algorithm 8. Extended Patterson Decoding

Input: A support set L, a Goppa polynomial $g(X)$ and a syndrome vector $s \in \mathbb{F}_2^{n-k}$
Output: A vector $e \in \mathbb{F}_2^n$ with $w_H(e) = t$
1: **procedure** RECERRVEC($L, g(X); s$)
2:    $s(X) \leftarrow$ TOPOLY(s)
3:    $v(X), g_1(X), g_2(X), g_{12}(X) \leftarrow$ CONSTRUCTKEYEQN($s(X), g(X)$)
4:    $a_2(X), b_1(X) \leftarrow$ SOLVEKEYEQN($v(X), g_{12}(X), \lfloor \frac{t}{2} \rfloor - \deg(g_2), \lfloor \frac{t-1}{2} \rfloor - \deg(g_1)$)
5:    $a(X), b(X) \leftarrow a_2(X)g_2(X), b_1(X)g_1(X)$
6:    $\sigma(X) \leftarrow a^2(X) + b^2(X)X$
7:    $e \leftarrow$ FINDERRVEC($\sigma(X)$)
8:    **return** e
9: **end procedure**

Input: A syndrome vector $s = (s_0, s_1, \ldots, s_{13t-1}) \in \mathbb{F}_2^{13t}$
Output: A syndrom polynomial $s(X) \in \mathbb{F}_{2^{13}}[X]$
1: **procedure** TOPOLY(s)
2:    $w_j \leftarrow \sum_{i=0}^{12} s_{13j+i} z^i \in \mathbb{F}_{2^{13}}$ for $j = 0, 1, \ldots, t-1$
3:    $s(X) \leftarrow \sum_{j=0}^{t-1} w_j X^j \in \mathbb{F}_{2^{13}}[X]$
4:    **return** $s(X)$
5: **end procedure**

Output: A syndrome polynomial $s(X)$ and a Goppa polynomial $g(X)$
Input: $v(X), g_1(X), g_2(X), g_{12}(X) \in \mathbb{F}_{2^{13}}[X]$
1: **procedure** CONSTRUCTKEYEQN($s(X), g(X)$)
2:    $\widetilde{s}(X) \leftarrow 1 + Xs(X)$
3:    $g_1(X), g_2(X) \leftarrow \gcd(g(X), s(X)), \gcd(g(X), \widetilde{s}(X))$ ▷ $g_1(X), g_2(X)$ are monic.
4:    $g_{12}(X) \leftarrow \frac{g(X)}{g_1(X)g_2(X)}$
5:    $\widetilde{s}_2(X), s_1(X) \leftarrow \frac{\widetilde{s}(X)}{g_2(X)}, \frac{s(X)}{g_1(X)}$
6:    $u(X) \leftarrow g_1(X)\widetilde{s}_2(X)(g_2(X)s_1(X))^{-1} \bmod g_{12}(X)$
7:    $v(X) \leftarrow \sqrt{u(X)} \bmod g_{12}(X)$
8:    **return** $v(X), g_1(X), g_2(X), g_{12}(X)$
9: **end procedure**

Output: $v(X), g_{12}(X), dega, degb$
Input: $a_1(X), b_2(X)$ s.t. $b_2(X)v(X) \equiv a_1(X) \pmod{g_{12}(X)}$ and $\deg(a_1) \leq dega, \deg(b_2) \leq degb$
1: **procedure** SOLVEKEYEQN($v(X), g_{12}(X), dega, degb$)
2:    $a_0(X), a_1(X) \leftarrow v(X), g_{12}(X)$
3:    $b_0(X), b_1(X) \leftarrow 1, 0$
4:    **while** $a_1(X) = 0$ **do**
5:        $q(X), r(X) \leftarrow div(a_0(X), a_1(X))$ ▷ $a_0(X) = a_1(X)q(X) + r(X), 0 \leq \deg(r) < \deg(a_1)$
6:        $a_0(X), a_1(X) \leftarrow a_1(X), r(X)$
7:        $b_2(X) \leftarrow b_0(X) - q(X)b_1(X)$
8:        $b_0(X), b_1(X) \leftarrow b_1(X), b_2(X)$
9:        **if** $\deg(a_0) \leq dega$ and $\deg(b_0) \leq degb$ **then**
10:            **break**
11:        **end if**
12:    **end while**
13:    **return** $a_0(X), b_0(X)$
14: **end procedure**

Output: An error locator polynomial $\sigma(X)$ and a support set L
Input: An error vector $e \in \mathbb{F}_2^n$
1: **procedure** FINDERRVEC(σ, L)
2:    $e = (e_0, \ldots, e_{n-1}) \leftarrow (0, 0, \ldots, 0)$
3:    **for** $j = 0$ to $n-1$ **do**
4:        **if** $\sigma(\alpha_j) = 0$ **then**
5:            $e_j \leftarrow 1$
6:        **end if**
7:    **end for**
8:    **return** e
9: **end procedure**

Algorithm 9. ENCRYPT, DECRYPT, ENCAP, DECAP, GETERRVEC, RO_G, RO_H

Input: A public key $pk = \widehat{\mathbf{H}}_{[n-k:n]} \in \mathbb{F}_2^{(n-k)\times k}$ and a vector $\widehat{e} \in \mathbb{F}_2^n$ with $w_H(\widehat{e}) = t$
Output: A syndrome vector $\widehat{s} \in \mathbb{F}_2^{n-k}$

procedure ENCRYPT($pk = \widehat{\mathbf{H}}_{[n-k:n]}; \widehat{e}$)
  $\widehat{\mathbf{H}} \leftarrow [\mathbf{I}_{n-k} \mid \widehat{\mathbf{H}}_{[n-k:n]}] \in \mathbb{F}_2^{(n-k)\times n}$
  $\widehat{s} \leftarrow \widehat{\mathbf{H}}\widehat{e} \in \mathbb{F}_2^{n-k}$
  return $\widehat{s}$
end procedure

Input: A secret key $sk = (L, g(X), \mathbf{S}^{-1}, r)$ and a syndrome vector $\widehat{s} \in \mathbb{F}_2^{n-k}$
Output: A vector $\widehat{e} \in \mathbb{F}_2^n$ with $w_H(\widehat{e}) = t$

procedure DECRYPT($sk = (L, g(X), \mathbf{S}^{-1}, r); \widehat{s}$)
  $s \leftarrow \mathbf{S}^{-1}\widehat{s}$
  $e \leftarrow$ RECERRVEC($L, g(X); s$) ▷ Alg. 8
  $\mathbf{P}, \mathbf{P}^{-1} \leftarrow$ GETPERMMAT(r) ▷ Alg. 6
  $\widehat{e} \leftarrow \mathbf{P}^{-1}e$
  return $\widehat{e}$
end procedure

Input: A public key $pk \in \{0,1\}^{(n-k)\times n}$
Output: A key $\kappa \in \{0,1\}^{256}$ and a ciphertext $c = (\widehat{r}, \widehat{s}) \in \{0,1\}^{256} \times \{0,1\}^{n-k}$

procedure ENCAP(pk)
  $r^* \xleftarrow{\$} \{0,1\}^{256}$
  $e^* \leftarrow$ GETERRVEC(r^*) ▷ $w_H(e^*) = t$
  $\widehat{r} \leftarrow \text{RO}_G(e^*)$ ▷ $\widehat{r} \in \{0,1\}^{256}$
  $\mathbf{P}, \mathbf{P}^{-1} \leftarrow$ GETPERMMAT($\widehat{r}$)
  $\widehat{e} \leftarrow \mathbf{P}e^*$
  $\widehat{s} \leftarrow$ ENCRYPT($pk; \widehat{e}$) ▷ $\widehat{s} \in \{0,1\}^{n-k}$
  $\kappa \leftarrow \text{RO}_H(e^* \| \widehat{r} \| \widehat{s})$ ▷ $\kappa \in \{0,1\}^{256}$
  return κ and $c = (\widehat{r}, \widehat{s})$
end procedure

Input: A secret key $sk = (L, g(X), \mathbf{S}^{-1}, r)$ and a ciphertext $c = (\widehat{r}, \widehat{s})$
Output: A key $\kappa \in \{0,1\}^{256}$

procedure DECAP($sk = (L, g(X), \mathbf{S}^{-1}, r); c = (\widehat{r}, \widehat{s})$)
  $\widehat{e} \leftarrow$ DECRYPT($sk; \widehat{s}$)
  $\mathbf{P}, \mathbf{P}^{-1} \leftarrow$ GETPERMMAT($\widehat{r}$)
  $e^* \leftarrow \mathbf{P}^{-1}\widehat{e}$ ▷ $\widehat{e}, e^* \in \{0,1\}^n$
  $\widehat{r}' \leftarrow \text{RO}_G(e^*)$
  $\widetilde{e} \leftarrow$ GETERRVEC(r)
  if $\widehat{r}' \neq \widehat{r}$ **then**
    $\kappa \leftarrow \text{RO}_H(\widetilde{e} \| \widehat{r} \| \widehat{s})$ ▷ implicit rejection
  else
    $\kappa \leftarrow \text{RO}_H(e^* \| \widehat{r} \| \widehat{s})$
  end if
  return κ
end procedure

Input: A 256-bit seed $r \in \{0,1\}^{256}$
Output: An error vector $e = (e_0, e_1, \ldots, e_{n-1}) \in \mathbb{F}_2^n$ with $w_H(e) = t$

procedure GETERRVEC(r)
  $e = (e_0, e_1, \ldots, e_{n-1}) \leftarrow (0, 0, \ldots, 0)$
  $[l_0, l_1, \ldots, l_{n-1}] \leftarrow$ SHUFFLE($[0:n], r$)
  for $j = 0$ to $t-1$ **do**
    $e_{l_j} \leftarrow 1$
  end for
  return e
end procedure

Input: An l-bit string $x \in \{0,1\}^l$
Output: a 256-bit string $r \in \{0,1\}^{256}$

procedure $\text{RO}_G(x)$
  return **LSH**("PALOMAGG"$\|x)_{[0:256]}$
end procedure

procedure $\text{RO}_H(x)$
  return **LSH**("PALOMAHH"$\|x)_{[0:256]}$
end procedure

References

1. Becker, A., Joux, A., May, A., Meurer, A.: Decoding random binary linear codes in $2^{n/20}$: How 1+1 Improves Information Set Decoding. In: Pointcheval, D., Johansson, T. (eds.) EUROCRYPT 2012. LNCS, vol. 7237, pp. 520–536. Springer, Heidelberg (2012). https://doi.org/10.1007/978-3-642-29011-4_31
2. Berlekamp, E.: Nonbinary bch decoding (abstr.). IEEE Trans. Inf. Theory **14**(2), 242–242 (1968)
3. Berlekamp, E., McEliece, R., van Tilborg, H.: On the inherent intractability of certain coding problems (corresp.). IEEE Trans. Inf. Theory **24**(3), 384–386 (1978)
4. Bernstein, D., et al.: Classic mceliece (2017)
5. Bernstein, D.J., Lange, T., Peters, C.: Attacking and defending the McEliece cryptosystem. In: Buchmann, J., Ding, J. (eds.) PQCrypto 2008. LNCS, vol. 5299, pp. 31–46. Springer, Heidelberg (2008). https://doi.org/10.1007/978-3-540-88403-3_3
6. Bernstein, D.J., Lange, T., Peters, C.: Smaller decoding exponents: ball-collision decoding. In: Rogaway, P. (ed.) CRYPTO 2011. LNCS, vol. 6841, pp. 743–760. Springer, Heidelberg (2011). https://doi.org/10.1007/978-3-642-22792-9_42
7. Bezzateev, S.V., Noskov, I.K.: Patterson algorithm for decoding separable binary goppa codes. In: 2019 Wave Electronics and its Application in Information and Telecommunication Systems (WECONF), pp. 1–5 (2019)
8. Bezzateev, S., Shekhunova, N.: Totally decomposed cumulative goppa codes with improved estimations. Designs, Codes and Cryptography 87(2), March 2019
9. Canteaut, A., Chabanne, H., national de recherche en informatique et en automatique (France). Unité de recherche Rocquencourt, I.: A Further Improvement of the Work Factor in an Attempt at Breaking McEliece's Cryptosystem. Rapports de recherche, Institut national de recherche en informatique et en automatique (1994)
10. Faugère, J.C., Gauthier-Umanã, V., Otmani, A., Perret, L., Tillich, J.P.: A distinguisher for high rate mceliece cryptosystems. In: 2011 IEEE Information Theory Workshop, pp. 282–286 (2011)
11. Finiasz, M., Sendrier, N.: Security bounds for the design of code-based cryptosystems. In: Matsui, M. (ed.) ASIACRYPT 2009. LNCS, vol. 5912, pp. 88–105. Springer, Heidelberg (2009). https://doi.org/10.1007/978-3-642-10366-7_6
12. Fujisaki, E., Okamoto, T.: Secure integration of asymmetric and symmetric encryption schemes. In: Wiener, M. (ed.) CRYPTO 1999. LNCS, vol. 1666, pp. 537–554. Springer, Heidelberg (1999). https://doi.org/10.1007/3-540-48405-1_34
13. Goppa, V.D.: A new class of linear error-correcting codes. Probl. Inf. Transm. **6**, 300–304 (1970)
14. Hofheinz, D., Hövelmanns, K., Kiltz, E.: A modular analysis of the fujisaki-okamoto transformation. In: Kalai, Y., Reyzin, L. (eds.) Theory of Cryptography, pp. 341–371. Springer, Cham (2017)
15. Karp, R.M.: Reducibility among Combinatorial Problems, pp. 85–103. Springer, US, Boston, MA (1972)
16. Kim, D.-C., Hong, D., Lee, J.-K., Kim, W.-H., Kwon, D.: LSH: a new fast secure hash function family. In: Lee, J., Kim, J. (eds.) ICISC 2014. LNCS, vol. 8949, pp. 286–313. Springer, Cham (2015). https://doi.org/10.1007/978-3-319-15943-0_18
17. Lee, P.J., Brickell, E.F.: An observation on the security of McEliece's public-key cryptosystem. In: Barstow, D., Brauer, W., Brinch Hansen, P., Gries, D., Luckham, D., Moler, C., Pnueli, A., Seegmüller, G., Stoer, J., Wirth, N., Günther, C.G. (eds.) EUROCRYPT 1988. LNCS, vol. 330, pp. 275–280. Springer, Heidelberg (1988). https://doi.org/10.1007/3-540-45961-8_25

18. Leon, J.S.: A probabilistic algorithm for computing minimum weights of large error-correcting codes. IEEE Trans. Inf. Theory **34**(5), 1354–1359 (1988)
19. Massey, J.: Shift-register synthesis and bch decoding. IEEE Trans. Inf. Theory **15**(1), 122–127 (1969)
20. May, A., Meurer, A., Thomae, E.: Decoding random linear codes in $o(2^{0.054n})$. In: Lee, D.H., Wang, X. (eds.) Advances in Cryptology - ASIACRYPT 2011, pp. 107–124. Springer, Heidelberg (2011)
21. May, A., Ozerov, I.: On computing nearest neighbors with applications to decoding of binary linear codes. In: Oswald, E., Fischlin, M. (eds.) EUROCRYPT 2015. LNCS, vol. 9056, pp. 203–228. Springer, Heidelberg (2015). https://doi.org/10.1007/978-3-662-46800-5_9
22. McEliece, R.J.: A public-key cryptosystem based on algebraic coding theory. Deep Space Network Progress Report **44**, 114–116 (1978)
23. Minder, L., Shokrollahi, A.: Cryptanalysis of the sidelnikov cryptosystem. In: Naor, M. (ed.) Advances in Cryptology - EUROCRYPT 2007, pp. 347–360. Springer, Heidelberg (2007)
24. Niederreiter, H.: Knapsack-type cryptosystems and algebraic coding theory. In: Problems of Control and Information Theory 15, pp. 159–166 (1986)
25. Patterson, N.: The algebraic decoding of goppa codes. IEEE Trans. Inf. Theor. **21**(2), 203–207 (2006)
26. Prange, E.: The use of information sets in decoding cyclic codes. IRE Trans. Inf. Theory **8**(5), 5–9 (1962)
27. Sidelnikov, V.M., Shestakov, S.O.: On insecurity of cryptosystems based on generalized reed-solomon codes. Discret. Math. Appl. **2**(4), 439–444 (1992)
28. Stern, J.: A method for finding codewords of small weight. In: Cohen, G., Wolfmann, J. (eds.) Coding Theory 1988. LNCS, vol. 388, pp. 106–113. Springer, Heidelberg (1989). https://doi.org/10.1007/BFb0019850
29. Targhi, E.E., Unruh, D.: Post-quantum security of the Fujisaki-Okamoto and OAEP Transforms. In: Hirt, M., Smith, A. (eds.) TCC 2016. LNCS, vol. 9986, pp. 192–216. Springer, Heidelberg (2016). https://doi.org/10.1007/978-3-662-53644-5_8

Author Index

A. Esser and P. Santini (Eds.): CBCrypto 2023, LNCS 14311, p. 175, 2023.
https://doi.org/10.1007/978-3-031-46495-9

Printed in the United States
by Baker & Taylor Publisher Services